D0747160

Communication and Public Participation in Environmental Decision Making

SUNY series in Communication Studies
Dudley D. Cahn, editor

Communication and Public Participation in Environmental Decision Making

Edited by

Stephen P. Depoe
John W. Delicath
Marie-France Aepli Elsenbeer

State University of New York Press

Published by
State University of New York Press, Albany

© 2004 State University of New York

All rights reserved

Printed in the United States of America

No part of this book may be used or reproduced in any manner whatsoever without prior written permission. No part of this book may be stored in a retrieval system or transmitted in any form or by any means including electronic, electrostatic, magnetic tape, mechanical, photocopying, recording, or otherwise without the prior permission in writing of the publisher.

For information, address State University of New York Press,
90 State Street, Suite 700, Albany, NY 12207

Production by Kelli Williams
Marketing by Fran Keneston

Library of Congress Cataloging-in-Publication Data

Communication and public participation in environmental decision making /
edited by Stephen P. Depoe, John W. Delicath, Marie-France Aepli Elsenbeer.
p. cm. — (SUNY series in comunication studies)
Includes bibliographical references and index.
ISBN 0-7914-6023-1 (alk. paper)
1. Environmental policy—Decision making—Citizen participation.
2. Environmentalism. I. Depoe, Stephen P., 1959– II. Delicath, John W.
III. Elsenbeer, Marie-France Aepli. IV. Series.

GE 170.C6413 2004
363.7'0525'0973—dc22

2004041627

10 9 8 7 6 5 4 3 2 1

Contents

List of Figures and Tables ix

Acknowledgments xi

Introduction 1
Stephen P. Depoe and John W. Delicath

Part One. Theorizing and Constructing More Effective Public Participation Processes

1. The Trinity of Voice: The Role of Practical Theory in Planning and Evaluating the Effectiveness of Environmental Participatory Processes 13
Susan L. Senecah

2. A Social Communication Perspective Toward Public Participation: The Case of the Cispus Adaptive Management Area 35
Amanda C. Graham

3. Competing and Converging Values of Public Participation: A Case Study of Participant Views in Department of Energy Nuclear Weapons Cleanup 59
Jennifer Duffield Hamilton

4. Public Expertise: A Foundation for Citizen Participation in Energy and Environmental Decisions 83
William J. Kinsella

Part Two. Evaluating Mechanisms for Public Participation in Environmental Decision Making

5. Decide, Announce, Defend: Turning the NEPA Process into an Advocacy Tool Rather than a Decision-Making Tool 99
Judith Hendry

6. The Roadless Areas Initiative as National Policy: Is Public Participation an Oxymoron? 113
Gregg B. Walker

7. Public Participation and (Failed) Legitimation: The Case of Forest Service Rhetorics in the Boundary Waters Canoe Area 137
Steve Schwarze

8. Public Involvement, Civic Discovery, and the Formation of Environmental Policy: A Comparative Analysis of the Fernald Citizens Task Force and the Fernald Health Effects Subcommittee 157
Stephen P. Depoe

9. Public Participation or Stakeholder Frustration: An Analysis of Consensus-Based Participation in the Georgia Ports Authority's Stakeholder Evaluation Group 175
Caitlin Wills Toker

10. "Free Trade" and the Eclipse of Civil Society: Barriers to Transparency and Public Participation in NAFTA and the Free Trade Area of the Americas 201
J. Robert Cox

Part Three. Emergent Participation Practices Among Activist Communities

11. Global Governance and Social Capital: Mapping NGO Capacities in Different Institutional Contexts 223
Amos Tevelow

12. Toxic Tours: Communicating the "Presence" of Chemical Contamination 235
Phaedra C. Pezzullo

13. Art and Advocacy: Citizen Participation Through Cultural Activism 255
John W. Delicath

Bibliography 267

List of Contributors 301

Index 305

Figures and Tables

Figures

6.1 Specific Directions in NEPA for Involving the Public 116

6.2 Core Values and Guiding Principles for the Practice of Public Participation 118

6.3 The IAP2 Public Participation Spectrum 119

Tables

3.1 Summary of Differences in Perspectives 65

6.1 Comparing Traditional and Innovative Public Participation 120

6.2 From President Clinton to the Final Rule 127

8.1 FCTF vs. FHES: Direct Participation of Diverse Perspectives 164

8.2 FCTF vs. FHES: Access to Information 165

8.3 FCTF vs. FHES: Person-to-Person Discussions 167

8.4 FCTF vs. FHES: Distribution of Power 169

Acknowledgments

This book sprang from many sources. Its initial impetus was the Sixth Biennial Conference on Communication and Environment hosted by the University of Cincinnati's Center for Environmental Studies in the summer of 2001. The conference was attended by nearly one hundred scholars, community members, and government officials from the United States and a number of other countries, including Australia, the United Kingdom, Canada, Brazil, Argentina, and Sweden. The conference theme was "Environmental Communication and Public Participation in Environmental Decision-Making." Those in attendance presented papers and shared a variety of hands-on experiences, including taking "toxic tours" of the Cincinnati communities of Lower Price Hill and Winton Place. It was an enlightening, transforming weekend for many of us.

The conference, and the book project that resulted, received a tremendous amount of support from a number of organizations that deserve mention here. We want to thank our major sponsors, including Scripps Howard Foundation, the Radway Environmental Information Project, the University of Cincinnati's Institute for Community Partnerships, the University of Cincinnati's Just Community Initiative, and the Department of African-American Studies. We also wish to acknowledge the Sierra Club's Environmental Justice Special Projects Grant Program for providing support for the "toxic tours" that took place during the conference. Additionally, we would like to thank Ms. Tammy Cromer-Campbell for bringing her exhibition of photographs entitled "Fruit of the Orchard: Pollution, Environmental Justice, and Social Responsibility" to the conference.

Finally, we wish to express our appreciation to the faculty, staff, and students of the University of Cincinnati's Department of Communication for lending their moral and material support to the project.

Introduction

Stephen P. Depoe and John W. Delicath

For the past several decades, interested citizens, policymakers, and regulators in the United States and elsewhere have asked: What role should community members play in the development and implementation of environmental policies? How can participation formats and processes be improved to maximize the value and impact of public input in environmental decision making? This book attempts to enhance our understanding of how community members and other interested parties engage in various kinds of participation, both within and outside institutionally prescribed formats, to influence environmental policy decisions. More specifically, chapters included in this volume are grounded in the assumption that issues of communication play a central role in questions related to effective public participation in environmental decision making. We hope that the communication theories and practices explored here can lead to improvements in the design, implementation, and assessment of both traditional and innovative public-participation mechanisms and formats.

The Issue of Public Participation in Environmental Decision Making

Since the passage of the National Environmental Policy Act (NEPA) in 1969, public participation in environmental decision making in the United States has become gradually institutionalized at federal, state, and local levels. The movement to involve citizens in environmental policy has spread to other countries as well through international forums such as the United Nations and the World Bank. At the same time, controversies over a wide variety of environmental issues, including facility siting, permit granting, natural resource management, land use, environmental justice, brownfields revitalization, smart growth, and international trade agreements, have led to an increasing focus on matters of public participation. Citizens, activists, and advocacy organizations in the United States and elsewhere have discovered firsthand the shortcomings of contemporary approaches to and mechanisms for citizen involvement, and have demanded changes in the way public participation is solicited and used.

More recently, public participation practitioners, citizens, and academics alike have begun to seek ways to promote more meaningful citizen involvement in environmental decisions. In some cases, government institutions responsible for providing for public involvement in environmental policymaking have explored changes to address the limitations of existing approaches. In the United States, for example, agencies such as the Environmental Protection Agency (EPA), the Department of Energy, the Forest Service, and the Bureau of Land Management have experimented with new forms of citizen involvement such as community-based collaborations, community advisory boards, citizen review panels, and on-line discussion forums as ways to develop mechanisms to improve public participation in environmental decisions.

A growing number of scholars and activists, and other interested citizens, have focused attention on the role of public participation in environmental decision making (Arnstein, 1969; Chess & Purcell, 1999; Daniels & Walker, 2001; Delicath, 2001; Fiorino, 1990, 1996; Kaminstein, 1996; Kasperson, 1986; Lynn & Busenberg, 1995; Renn, Webler, & Wiedemann, 1995; Rosenbaum, 1978; Tuler & Webler, 1995; Webler, Tuler, & Krueger, 2001). This extensive literature has identified a number of shortcomings in traditional participation mechanisms (such as public hearings or comment periods), including:

(1) Public participation typically operates on technocratic models of rationality in which policymakers, administrative officials, and experts see their role as one of educating and persuading the public about the legitimacy of their decisions.

(2) Public participation often occurs too late in the decision-making process, sometimes even after decisions have already been made.

(3) Public participation often follows an adversarial trajectory, especially when public participation processes are conducted in a "decide-announce-defend" mode on the part of officials.

(4) Public participation often lacks adequate mechanisms and forums for informed dialogue among stakeholders.

(5) Public participation often lacks adequate provisions to ensure that input gained through public participation makes a real impact on decision outcomes.

Along with criticizing traditional mechanisms, many scholars and practitioners have attempted to articulate a set of assumptions and values, grounded in ideals of participatory democracy, upon which alternative public participation processes can be developed and evaluated (Delicath, 2001; Fiorino, 1989a, 1989b; Laird, 1993; Rosenbaum, 1978). Principal among these assumptions are the following: (1) people should have a say in decisions that will affect their lives; (2) early and ongoing, informed and empowered public participation is the hallmark of sound public policy; and (3) the public must be involved in determining how they will participate in choosing what forums and mechanisms will be used in identifying what resources are needed to ensure informed participation, and in determining how public input will affect decision-making outcomes. These assumptions have led a number of scholars to outline approaches to more democratic public participation in environmental decision making based on notions of fairness and competence (Renn, Webler, & Wiedemann, 1995), collaborative learning (Daniels & Walker, 2001) and rhetorical models of risk communication (Katz & Miller, 1996; Rowan, 1994; Waddell, 1996). Still others have examined specific mechanisms that attempt to give the public a larger role in environmental policy decision making, such as community advisory boards, citizen panels, citizen advisory committees, and citizen juries (Applegate, 1998; Crosby, 1995; Goldenberg & Frideres, 1986; Vari, 1995).

Although there is growing agreement about the assumptions and values of democratic participation, the shortcomings of contemporary approaches to public participation, and the general types of changes necessary to achieve more meaningful citizen involvement in environmental decision making, discussions of these issues have paid inadequate attention to issues of communication. One of the main objectives of this book is to highlight the centrality of communication in matters of public participation in environmental decision making.

COMMUNICATION APPROACHES TO PUBLIC PARTICIPATION

Chapters in this volume are grounded in environmental communication studies, an emerging research tradition that explores the ways in which communication—strategic symbolic action shared among people and organizations—impacts both our conception of and our interaction with the physical world (Cantrill & Oravec, 1996; DeLuca, 1999; Herndl & Brown, 1996; Killingsworth & Palmer, 1992; Muir & Veenendall, 1996; Peterson, 1997). Scholars within this tradition have noted that environmental communication serves at least two important functions, both of which are relevant to issues of public participation in environmental decision making. First, environmental communication serves an instrumental function. As human beings make daily decisions, as individuals and in communities, about how they intend to manage and care for the physical spaces around them, those decisions are greatly influenced by advocacy for various environmental values, priorities, and policies. Hence, environmental communication scholarship offers critical analyses of the persuasive efforts of advocates from across the political spectrum, including individuals, grassroots organizations, corporations, and government agencies, who attempt to shape decision-making processes and policy outcomes. This practical function of environmental communication is clearly relevant to the study of public participation mechanisms and practices. Participants in environmental decision making utilize strategic communication in efforts to set agendas, define problems, and advocate solutions, as well as to cultivate trust, articulate community voice and vision, and build legitimacy for decisions.

Second, environmental communication serves a constitutive function. In many respects, "the environment" is not simply a material object or site out there beyond the individual, but is also a symbolic construct created and organized through discourse. Environmental communication scholarship raises ontological issues related to tensions between conceptions of a nature that is physically experienced as material substance and a "Nature" that is symbolically constructed and enacted through discourse (Herndl & Brown, 1996). This function of environmental communication is particularly relevant to public participation. All participants in environmental decision making communicate in ways that not only represent problems, causes, solutions, legitimacy, interests, and values, but also construct those very issues in question. Indeed, as Fischer and Forester point out, public participation in environmental decision making is "a constant discursive struggle over the criteria of social classification, the boundaries of problem categories, the intersubjective interpretation of common experiences, the conceptual framing of problems, and the definitions of ideas that guide the ways people create the shared meanings which motivate them to act" (1993, pp. 1–2).

The recognition that our communication about environmental matters serves both instrumental and constitutive functions provides the theoretical and critical foundation of this volume, and allows for a more thorough examination of the dynamics of stakeholder involvement in deliberations about environmental issues. Operating from various approaches to the study of communication, authors in this volume come to a number of conclusions that add to—and at times cast doubt on—the accepted wisdom about the purposes, structures, and outcomes of public participation in environmental decision making.

ORGANIZATION OF THE VOLUME

The chapters in this book are organized in three parts. Part 1, "Theorizing and Constructing More Effective Public Participation Processes," explores the role of communication theory in designing, executing, and evaluating mechanisms for effective public participation in environmental decision making. Drawing on literature from group communication, Susan L. Senecah outlines a practical theory involving what she calls the "trinity of voice," and argues for its utility as a tool for assessing the effectiveness of participatory processes. Amanda C. Graham presents a social communication framework, based on values of openness, shared responsibility, and interpersonal relationships, as an alternative model that opens up the possibility for enhanced engagement among stakeholders in environmental decision-making contexts. Working from a "competing values" theory of group decision making, Jennifer Duffield Hamilton explores the extent to which participants' expectations about the purpose, structure, and outcomes of public participation may influence the effectiveness of participatory mechanisms. Finally, William J. Kinsella problematizes distinctions between "experts" and "the public" that are common in traditional public participation practices and scholarship, with the goal of identifying both how community members can become more capable of evaluating expert arguments and how expert knowledge should fit within the larger framework of environmental policymaking. Kinsella argues that "expertise" should be conceived as a "public resource"—a kind of social knowledge that accounts for both expert and local knowledge, and to which both specialists and non-specialists contribute.

Part 2, "Evaluating Mechanisms for Public Participation in Environmental Decision Making," takes a critical look at public participation practices and processes within current institutional frameworks in the United States and abroad. The chapters in this section offer case studies of local and national public participation processes in institutional contexts such as NEPA, the U.S. EPA, the U.S. Forest Service, and the Department of Energy. International contexts of public participation, such as regional trade agreements such as the North American Free Trade Agreement (NAFTA), the Free Trade Areas of the Americas (FTAA), the World Trade Organization (WTO), the World Bank (WB), and the United Nations are examined as well.

Judith Hendry offers a specific critique of public participation requirements within the National Environmental Policy Act (NEPA), with a particular focus on how the Environmental Assessment segment of the law operates as a de facto advocacy tool to validate a priori decisions rather than as a decision-making tool to arrive at carefully weighed decisions. The next two chapters examine public participation practices in the U.S. Forest Service (USFS). Gregg B. Walker assesses the efforts by the Clinton administration to gather public comments as part of the USFS implementation of the 1999 Roadless Initiative, a plan that called for the protection of nearly 40 million acres of roadless areas throughout the U.S. national forest system. Walker's chapter reveals that despite institutional commitments to civic engagement and a concerted effort to solicit public input, the Roadless Initiative public participation strategy exemplified a "business as usual" approach rather than innovation and civic deliberation. Steve Schwarze's chapter looks at the USFS management plans developed for the Boundary Waters Canoe Area Wilderness in northeastern Minnesota and finds that the agency's public participation efforts resulted in damage to its own legitimacy. Schwarze argues that the bureaucratic and instrumental rationality that guides the Forest Service's solicitation of public participation and its account of public participation as reported in USFS plans creates an "institutional trap" that erodes stakeholder confidence and trust in the organization. The next pair of chapters examines structures and practices related to citizen review panels. Stephen P. Depoe argues that the effectiveness of the Fernald Citizens Task Force and the Fernald Health Effects Subcommittee, panels established by the Department of Energy to provide consensus-based recommendations regarding environmental remediation and health research related to America's nuclear weapons complex, differed greatly in large part because of the institutional structures and tasks of the groups as well as because of how discursive practices within the panels were constrained by conventional models of expert knowledge and risk communication. After identifying problems associated with the Georgia Ports Authority's adoption of a consensus-based stakeholder approach to resolving conflicts associated with a plan to deepen Savannah's harbor, Caitlin Wills-Toker calls for abandoning not only consensus-based approaches, but other detached and abstract models of participation as well. Finally, J. Robert Cox examines two regional trade agreements—the North American Free Trade Agreement (NAFTA) and the proposed Free Trade Area of the Americas (FTAA)—to identify structures that resist transparency and the participation of civil society groups in trade negotiations and dispute-settlement procedures. Cox reveals an urgent need to consider public participation on regional and global scales and argues that the ability of civil society groups to participate in trade forums is dependent on articulating a compelling rationale for their inclusion and a coherent vision of alternative agreements that promote a just and sustainable economy.

Part 3, "Emergent Participation Practices Among Activist Communities," explores alternative, noninstitutional resources for and strategies of public participation in environmental decision making. In this section, public participation is broadly conceived and includes considerations of the roles of social capital, toxic tours, and cultural activism as novel means of citizen involvement in environmental policymaking. Following from the issues previously raised by Cox, Amos Tevelow examines the structural and rhetorical capacities of nongovernmental organizations (NGOs) to participate in international governance disputes surrounding issues of environment, agriculture, natural resources, disease, and human rights. Tevelow argues that NGO cooperation in a global policy network requires the cultivation of social capital in ways that can overcome major gaps in culture, resources, and power. Tevelow cautions, however, that social capital used as a policy tool may actually undermine more fundamental reforms by "humanizing" governance without rigorously and explicitly addressing issues of equity and justice. The final two chapters explore novel and innovative cultural practices of grassroots environmental justice organizations and their relationship to public participation in environmental decision making. Phaedra C. Pezzullo explores the persuasive dimensions of toxic tours as complex rhetorical strategies through which community-based "guides" or advocates provide a powerful critique against dominant discourses of toxic waste. Pezzullo explores how tours impact public participation by educating government representatives and creating opportunities for citizens to share strategies for responding to environmental injustice. Finally, John W. Delicath examines the role that cultural activism and photography play in the struggle for environmental justice. Delicath argues that participation theorists must consider the issues of what motivates, inspires, prepares, and empowers the public to participate in environmental decision making and examines the strategy of cultural activism as a means to explore the relationship between citizen advocacy in noninstitutionalized settings and public participation in institutionalized contexts.

CONTRIBUTIONS OF THE VOLUME

This book explores the communication practices of various stakeholders (citizens, grassroots organizations, advocacy groups, industry representatives, scientists and technical experts, federal regulators, government agencies) engaged in a variety of environmental decision-making contexts (natural-resource management, use of public lands, nuclear remediation, environmental justice, and world trade). Included are case studies that analyze individuals and organizations participating in a wide range of activities, both within institutional mechanisms (public hearings, NEPA processes, citizen advisory boards) and through alternative forms of environmental advocacy.

At one level, the chapters in this volume provide support for a number of commonly held conclusions concerning how public participation can be improved. It is clear that lawmakers, regulators, and others charged with developing and implementing environmental policies at local, national, and international levels need to articulate a clearer sense of decision space (where and when decisions are to be made) and decision authority (who makes the decision based on what factors) in ways that clarify how public participation will meaningfully impact environmental decisions.

At the same time, the book invites a more complex, critical examination of public participation that recognizes both the instrumental and constitutive functions of communication in environmental controversies. The studies in this volume reveal that public participation in environmental decision making is both shaped by and, in many cases, constrained by the ways in which environmental issues, problems, and solutions are defined or framed through the strategic communication practices of various stakeholders.

Contributors to this volume identify a number of cautionary notes regarding the possibilities and limitations of public participation in environmental decision making, including:

- The general principles of meaningful public participation consistently identified in the literature require constant operationalization. Researchers must always look to contextual factors when explaining why a particular mechanism or instance of participation was or was not successful.
- Providing structured opportunities for public input, including the use of innovative mechanisms to encourage participation, does not by itself guarantee meaningful citizen involvement leading to publicly supportable decisions.
- The meaning and value of public participation depends to a significant extent on how concepts such as "participation" and "participant" are defined by those involved in the process. Participants' expectations about the purpose, opportunities, structures, and outcomes of public participation, including their own potential to affect outcomes, will influence the effectiveness of participatory processes and the level of satisfaction with decisions.
- Efforts by policymakers, environmental advocates, and others to achieve meaningful public participation may be constrained by more deep-seated commitments to institutional rationalities or economic imperatives that are articulated in dominant discourses of expertise, knowledge, risk, and legitimacy.
- How environmental issues are spatialized at various levels of government (local, regional, national, international) and defined geographically, as affecting certain people and environments in particular places, structures and often constrains effective communication and meaningful public participation.
- Models or theories of meaningful public involvement in environmental decision making must account for the growing significance and impact of alternative modes of environmental advocacy, including toxic tours and cultural activism, as ways in which individuals gain the confidence and skills to participate in, and to transform, institutionalized processes of public participation.

In sum, the volume highlights ongoing tensions between philosophical calls for more democratic public participation in environmental decision making and the practices of institutions and public officials who too often acknowledge or solicit community input without adequately allowing for that input to influence policy choices or regulatory outcomes.

In applying communication theories to public participation in environmental decision making, we urge participants in environmental controversies to acknowledge the legitimacy of both technical expertise and local knowledge, and to seek a more productive dialogue among multiple discourses in which citizens, experts, and other participants articulate, interrogate, and transform each other's perspectives. An approach to public participation that accounts for both the instrumental and constitutive dimensions of environmental communication would recognize the contingent nature of knowledge and emphasize an interactive exchange of ideas in which all participants both communicate and appeal to facts, knowledge, and reason as well as beliefs, values, and emotions. Those who seek to improve public participation in environmental decision making should strive to develop mechanisms and forums of engagement that emphasize civic discovery, interpersonal relationships, collaborative learning, and deliberative rhetorics "through which citizens test and create social knowledge in order to uncover, assess, and resolve shared problems" (Goodnight, 1982, p. 214). We hope that this volume contributes to that effort.

Part One

Theorizing and Constructing More Effective Public Participation Processes

CHAPTER 1

THE TRINITY OF VOICE: THE ROLE OF PRACTICAL THEORY IN PLANNING AND EVALUATING THE EFFECTIVENESS OF ENVIRONMENTAL PARTICIPATORY PROCESSES

SUSAN L. SENECAH

Effectiveness in the communication processes by which a community makes its environmental decisions, especially contentious or potentially contentious decisions, is a key component of community sustainability, but it is not an easy component to achieve. This chapter suggests that we consider expanding Schutz's (1958) FIRO theory of inclusion, affection, and control needs of small group members to multi-stakeholder environmental policy decision-making processes such as public hearings, advisory groups, and task forces as well as to newer, emerging processes. The discussion introduces the practical theory of the Trinity of Voice (TOV)—access, standing, and influence—and discusses its usefulness as an effective benchmark against which to plan or evaluate participatory processes regarding contentious environmental issues. Discussion also addresses how the TOV can play an additional transformative role in building community capacity (Dukes, 1996). Finally, plenty of questions call to be answered, and I suggest several trajectories for future attention.

Because environmental decision making typically triggers competing demands for (1) specific, scientific/technical expertise; (2) ultimate decision making by governmental authorities; and (3) demands for inclusive public involvement, public participation processes, or the lack of them, are often flashpoints among competing perspectives resulting in the escalation of conflicts. This is not to suggest that all conflicts should be avoided. Certainly, social conflict is sometimes necessary to challenge social assumptions, raise awareness, and bring about necessary social change. The Civil Rights movement demonstrated this clearly. This is also not to suggest that all environmental conflicts should be avoided. For example, an official process that institutionalizes the marginalization of stakeholders may leave no alternative. Finally, sometimes there really is no space in an issue for creative problem solving or deliberation. For example, the answer might not necessarily be a smaller landfill or a smaller low-level radioactive waste-storage facility.

However, when conflict escalates at the local level, the anger and skepticism produced in the heat and aftermath persists far beyond the lawsuit, the rally on the statehouse lawn, the petitions, the fiery demonstrations at the public hearing, or any of the other demonstrative ways that stakeholders use to elbow their way through the walls of a thick process. Contentious local environmental conflict produces subtler, long-lasting residual effects that persist in communities and further erode their fragile covenant of trust with their elected officials, democracy, and each other. The effects of this civic disengagement have been well documented (DeWitt, 1994; Dunham, 1986; Edgerton, 1995; Healey, 1992; Keyes, 1973; Lyon, 1987; Putnam, 1995; Sandel, 1996; Sarason, 1974; Yankelovich, 1981) and although other sources contribute, ineffective public participatory processes can create a maelstrom of civic discontent.

However, from my decade of experience and the broad experience of colleagues, many local land use and other environmental decisions need not be contentious. Others may be contentious but need not be destructive. Effective participatory processes have the potential to not only support good environmental decision making, but build a community's ability to engage other issues in more productive ways that support a solid civic base and a higher quality of community experience and relationships. Therefore, if conflict can be avoided or minimized, and community capacity for engaging difficult, contentious issues can be increased through effective public participation, then increasing the chances that processes will be effective makes sense.

To that end, I offer a practical theory (Cronen, 1995, 2000) by which effectiveness of participatory processes can be assessed and their grammars of access, standing, and influence articulated. First, I briefly establish the background and challenges of effective public participation concerning environmental issues, followed by a general discussion of Cronen's notion of practical theory and the TOV. Finally, I discuss how the TOV can be used as a mechanism by which to (1) "front-end load" a decision-making process during the design phase, (2) evaluate either existing or ongoing processes, or (3) assess an escalated conflict for effectiveness.

BACKGROUND TO THE EVOLVING DEMANDS FOR PUBLIC PARTICIPATION

Traditionally, "the American character is marked by a passion for both democracy and expertise" (DeSario & Langton, 1987a, p. 205) and increasingly, it seems difficult to reconcile these two ideals. At least two kinds of expertise are prominent in public participation arenas regarding environmental decisions. One universal characteristic of current environmental problems is the ever-increasing level of specific scientific and technological knowledge required to understand, mitigate, and prevent environmental degradation. The other is the traditional decision-making process characterized by elected or appointed officials acting to represent their constituencies and to exercise their authority by making decisions for the public good.

Parallel to these experts' roles is the momentum that marks the contemporary desire on the part of citizens to have a meaningful role in determining the destiny of their communities. This is one of the more overt consequences of the complex political, cultural, and informational shifts occurring over at least the past four decades that have increased awareness of environmental matters and simultaneously raised expectations for more active public participation in governmental decisions bearing on the environment. Hence, institutionalized technocracy and participatory democracy have intersected as the main protagonists (DeSario & Langton, 1987a) in environmental decision making at the local level and is often marked by tension and even conflict.

Before the National Environmental Policy Act (NEPA) of 1970 mandated the preparation of and provision for public review of an environmental impact statement for any project that had the potential to cause negative environmental impacts, the generic public had little access to the formal decision-making process, nor did it expect much. Citizens were, by and large, seemingly content to be passive spectators who deferred to technocrats and elected authorities, especially on environmental matters. If they participated at all in civic life, it was most likely by voting.

However, as numerous commentaries and histories chronicle, the raised environmental consciousness and the political and cultural shifts that marked the 1960s and 1970s supported the emergence of other kinds of citizen and nongovernmental organizations (NGOs). They also compelled the recognition of multiple "publics" with competing perspectives and different expectations for involvement in environmental public policy decision making (DeSario & Langton, 1987a; Sirianni & Friedland, 1995). With the addition of a rising public distrust that government and industry were acting on behalf of the citizens' best interests (Peele, 1990), these citizens felt increasingly empowered to act individually, as members of NGOs, or as part of local citizen groups. As stakeholders in the outcome of the decision, these citizens felt compelled to elbow their way into arenas of democratic processes (Harless, 1992). Even with the introduction of the federal NEPA process and, consequently, individual state versions of it, the traditional decision-making bureaucracy was neither structured to nor altogether willing to accommodate these new expectations or behaviors (Harless, 1992).

As a result, most governing systems attempted to develop, and are now mandated by a litany of federal and state laws and executive orders to provide, a minimum of formal opportunities for public participation beyond voting. In addition, citizens might be invited to serve in local government (e.g., planning boards), appear and testify at legislative and administrative proceedings, or join commissions, task forces, advisory committees, or study groups. As a result, most stakeholders, administrators and citizens alike, can "talk the talk" of what characterizes effective public participation, for instance, openness, trust building, dialogue, feedback, active listening, and information flow. However, from my ten years of discussing this issue with hundreds of local officials and agency administrators, most, by their own admission, are either unwilling, incapable, or feel constrained to "walk the walk." In summary, from their vantage point, they

- believe that it would encourage the mobilization of antagonistic interests;
- are concerned that it will cost too much, consume time, and delay schedules;
- fear that the process will be taken over by special interests;
- fear that participants might not be truly representative;

- fear that it will undermine their authority or mandate as elected officials;
- fear that it will make them look incompetent, that they can't make good decisions on behalf of their community;
- believe that citizens lack capability or knowledge of complex, technical issues;
- are confused about how to weigh various forms of public input in making a decision;
- fear the loss of control over the process due to lack of confidence in process dynamics;
- view themselves as the elected representatives or technical experts (one administrator called this "professional elitism" and "technoarrogance");
- lack the skills and confidence required to interact effectively with diverse publics;
- lack managerial interest to design or engage nontraditional processes;
- perceive that the tactics and styles of citizen groups (based on experiences at public hearings and town board meetings) often seem overdramatized and hysterical. This supports a belief that the citizens' information or research is unsound. This leads to a belief that citizen concerns and demands are simply delaying tactics and are unjustified, resulting in a refusal to deal with "people like that."
- fear that complex issues become oversimplified. For example, economic disparities might be masked, policy options might be overlooked, and cost–benefit consciousness might be ignored.
- fear that any nontraditional activities will leave them legally exposed. Even those administrators who support more diverse and inclusive processes express hesitation to do so for this reason.

To a degree, I empathize with these officials. If my experiences with citizens at public hearings, for example, were characterized by angry, sign-waving, irrational-sounding, ignorant-of-the-facts, seemingly hysterical citizens, I'd probably not be so eager to let them into any more of the decision-making process. But these common experiences of officials and citizens alike beg a fundamental question. Are existing traditional public-participation processes the way they are because of the way publics act, or because of their apathy? Or do publics act the way they do because of the way the processes are? As I noted earlier, some conflicts are unavoidable and some are necessary. Nevertheless, what can be said about many local environmental disputes is that a significant incongruency exists between the expectations for public participation raised by the laws, executive orders, and treaties and the actual experiences of participants in those processes. This has led to several consequences.

One consequence has been that various understandings of what "effective" means have emerged. If your vantage point is administrative, effective certainly means that you fulfilled the agency's legal mandate and kept it out of court. Written notice was given in the correct publications at the correct interval. A forum was provided at the appropriate phase in the permitting process during which you transmitted public information or collected public comment. As Christopher Gates, president of the National Civic League puts it, "Yeah, did that. Yeah, did that. Can check that off the list" (Gates, 1998). As cynical as this sounds, "enduring the public gauntlet" is how the majority of administrative and industry views interfacing with publics in decision making about contentious environmental issues.

Providing information on risk and translating technical information into lay terms is a far more comfortable and predictable role, even if it is unpleasant. This is another understanding of what effective means. More often than not, according to Persons (1990) and many others, formal mechanisms for soliciting citizen input and participation generally tend to be ritualistic endeavors that at their worst are "designed [or are perceived] to shroud an elitist policy making process in the cloak of democracy" (p. 121) and perhaps, at best, are ineffective. At the very least, nearly everyone I have ever talked to agrees that the formalized, traditional forums for public participation are generally not effective in changing or improving a decision, nor are they effective or productive in improving community relationships, especially with elected officials. It is this latter point that is most often cited as the measure of effectiveness.

Another significant consequence of the incongruence between expectations and experience has been increased public frustration, disillusionment, skepticism, and anger as the traditional forums unfold (Shepherd & Bowler, 1997). As Christopher Gates (1998) notes, "[Citizens] get the joke" that their opinions or preferences don't count. The NIMBY (Not In My Backyard) phenomenon is one of the best-known responses to this (Mowrey & Redmond, 1993), but in addition, citizens have sought out legal redress, prompting industry to countersue citizens with what have come to be called Strategic Litigation Against Public Participation (SLAPP) suits. This, in turn, has prompted attempts to legislate against both citizens' ability to sue and SLAPPs. It would seem to be more productive to assess the process that tends to trigger this cycle of frustration.

A final significant consequence is that most stakeholders agree that much of this public involvement is not effective, not in terms of being productive or meaningful. An often-heard accusation is that the public had no voice. Therefore, whereas government agencies, NGOs, and industry have recognized, some reluctantly, that publics are entitled to real and meaningful (effective) participation, multiple tensions have emerged and continue to snarl from attempting to define and operationalize real and meaningful (Gariepy, 1991).

This produces a paradox. Even though the overwhelming majority of contemporary public administration literature extols the benefits and necessities of meaningful and early public participation; even though every U.S. agency produces manuals, training materials, and guidelines for effective public participation; and even though every international environmental treaty makes explicit provisions for effective public participation; DeSario and Langton's 1987 admonition still rings true: "We are far from clear on matters of what kinds of structures and procedures would be necessary to develop and maintain a healthy 'technodemocratic' society" (1987b, p. 14). What constitutes voice?

That is, we still are not quite sure what "effective" means, based on general principles that allow for case-specific accommodation but also serve to pull common threads through diverse cases (Popovic, 1993). Certainly, practitioners, administrators, and scholars have endeavored to identify the characteristics of effective processes. A plethora of early articles over the past three decades have critiqued public participation trends (for example, DeSario & Langton, 1987a; Persons, 1990; Rosenbaum, 1978; Wengert, 1976). Several extensive bibliographies have been compiled, and organizations, centers, and institutes founded to promote effective public participation have Web sites offering narratives along with guidance and training materials.[1] The growing body of case studies in the environmental conflict-resolution literature also chronicles the factors and incidents that escalate tensions to the seemingly intractable point at which a mediator enters the scene.

Although a small portion of the literature does focus on successful or effective processes for environmental decision making, most of the public-participation literature is a sort of backdoor lamenting and analyzing of what went wrong (e.g., Shepherd & Bowler [1997] analyze the U.S. Environmental Protection Agency's reliance on the ineffective public process). Studies repeatedly testify to stakeholder frustration with current available opportunities for participation and chronicle the destructive spiral of escalating conflict that often results (Mowrey & Redmond, 1993; Plant & Plant, 1992).

However, as informative as these narratives of effective and less effective processes are, and as sensible and useful the guidance that results from them, often they are too case specific or too generally vague in their analyses and conclusions. Further, empirical scholarly attention to assessing the effectiveness of even the most common structures for public participation is in its early stages and is conspicuous by its dearth. For some processes, it is conspicuous by its absence. For example, very little empirical research has addressed the actual processes themselves, especially the most popular, the public hearing. According to Chess (1995), calls for more academic research on this topic date back to the 1960s, yet there remains a significant gap. While somewhat informative, the identification of elements that particular processes did or did not provide (e.g., open, feedback, face-to-face, active listening and response, involving all stakeholders) is not particularly useful as a general theory on which we can consistently design, monitor, or remediate effectiveness across the continuum of public processes.

Here is a sampling of descriptors of effective public participation drawn from a diverse collection of organizational materials, agency and other training materials, scholarly articles, and citizen newsletters: accessibility, fairness, perceived understandability, empowerment, openness, consistency, dialogue early and often enough to keep stakeholders engaged, protection of minority rights and interests, improved decision making, even political playing field, even resource playing field, comprehensive representation, information flow, response, legitimacy, early involvement, dialogue and discussion, adequate time to talk, clarity about how the input will be utilized, conduciveness to collaboration. In short, we all seem to know effectiveness when we see or don't see it, but massaging these attributes into a general model that can be used by all regardless of individual contexts has proven to be more difficult.

What can be concluded from the anecdotes, the case studies, the training materials and guides, the government initiatives, and the limited scholarly research is that regardless of the myriad of descriptors that are used to describe effectiveness factors, in the end, it all leads to *trust* (for example, Harless, 1992; Lewicki & Wiethoff, 2000; Murray & Nobleza, 2001). Trust is overwhelmingly the most commonly identified missing or present element in ineffective or effective processes. It seems that all effective process elements, or the antithesis of ineffective process elements, ultimately support the creation, enhancement, or maintenance of trust.

However, how does one know what trust looks like? How does one know when it has been established? How can one front-end load a participatory process to best assure its integrity? How can one unravel an escalated conflict to determine where the raw wounds of distrust fester to resolve the conflict?

Trust is a difficult concept to operationalize, as evidenced by the previous list of process elements. We need a general theory that accounts for the elements of trust in public processes and allows the benchmarking of these components in designing, monitoring, or remediating public process. This general theory must also be useful and flexible in practice; it must be a practical theory.

PRACTICAL THEORY

Practical theory offers to link knowledge and practice. Cronen (1995) exhorts scholars to follow John Dewey's admonishment to develop ways of studying social phenomena that take human experience seriously, to value what is in process and fraught with possibility ("ends in view") as much as what is finished and perfected ("ends"), and to avoid taking socially constituted phenomena for natural facts or powers (pp. 220–221). "Meaning," as Cronen (1995) points out, "involves pointing beyond the moment" (p. 224). Think of the rules of the game of chess. Although chess players of all ranks know the basic rules of the game, what distinguishes a chess master from a player like me is a sophisticated knowledge of finessing the rules that allows a higher-level cognition and ability to strategically choose and adapt the rules in different game contexts. Same rules + a sixth sense savviness for finessing them + different game contexts = ultimate flexibility = higher likelihood of success. In the same way, practical theory guides the researcher and the practitioner in recognizing, developing, and enacting "flexible, expanding abilities that cannot be reduced to technique" (p. 224).

Cronen (1995) suggests features of practical theory that I take some liberty with in describing. They are:

- Patterns of practice that come to characterize a reality for participants;
- The way these practices can be noticed or in this case, encouraged—their grammars;

- A family of methods that helps the study of the action and that also respects the role that people play in constructing their own experience in the conjointly created reality;
- The provision of ways of joining in situated action that promote socially useful understandings of, and new abilities for, all parties involved (and I assume this includes the researcher).

Cronen (1995) insists that practical theory must lead to practices; "a practical theory's grammar of practice refers to the abilities a professional brings to a situation, joining with the abilities of others. These professional abilities are informed by a coherent way of going on—the practical theory" (p. 6). In essence, Cronen (2001) defines practical theory as follows: "A practical theory informs a grammar of practice that facilitates joining with the grammars of others to explore their unique patterns of situated action. . . . Practical theory itself is importantly informed by data created in the process of engagement with others" (p. 26).

THE TRINITY OF VOICE: ACCESS, STANDING, INFLUENCE

This practical theory plays off of Schutz's (1958) theory of fundamental interpersonal relations orientation (FIRO) that attempted to identify what members needed to achieve cohesiveness in small task groups. He argued that group members needed to feel a sense of inclusion (being noticed, having prestige and esteem), affection (close personal ties or a sense of being liked) and control (a sense of power). Although Bormann (1990) asserts that Schutz's approach was perhaps too sophisticated and his research questions too complex to be answered by the scaling devices available then, the needs of small-group members ring true for many practitioners. I wondered what his three universal small-group member needs would look like when expanded in scope and scale to multi-stakeholder, multiobjective needs in complex environmental public policy processes. I wondered if melding research and practice might not give us a practical theory with patterns of practice and grammars. The practical theory of the TOV could be thought of as sixth-sensed savviness and flexibility of knowing how and when to finesse the basic techniques that all public participation specialists should have in their toolboxes. The practical theory of the TOV (access, standing, influence) provides a rubric or a schematic that would support this sixth sense to design, assess, or remediate effectiveness in public processes.

THE INTERPLAY OF ACCESS, STANDING, AND INFLUENCE

The general TOV theory holds that the key to effective process is an ongoing relationship of trust building to enhance community cohesiveness and capacity, and results in good environmental decisions. This relationship may ebb and flow but the intention is always in the direction of preserving and enhancing trust. To paraphrase Aldo Leopold (1949/1987), that which tends to support the integrity of the community is good; that which does not is not good. The practices of access, standing, and influence must be present to build and maintain trust. Stakeholders who are denied any of these components will find a means by which to claim them. Often these means are disruptive, destructive, and unproductive in achieving decision effectiveness and social effectiveness.

Access

Access refers to opportunity, potential, and safety. In its simplest form, it means that I have access to sufficient and appropriate opportunities to express my choices and opinions, but it is more than this. It means that I have the opportunity to access sufficient and appropriate support, for instance, education, information, so that I can understand the process in an informed, active capacity, not as a reactionary. These opportunities may be direct, indirect, or vicarious (e.g., via local-access cable channels). It means that I am in a space that holds the potential for me to be heard. I like the way Popovic (1993) characterizes access:

> Educated and equipped with information, the concerned person . . . needs a forum in which to express his or her concerns—a place to make something happen, to step into the decision-making process, or in some cases, a way to get a decision-making process started. Public input in the decisional process includes suffrage, but it goes much farther. The heart of public voice is the right of free expression, supported by the rights of assembly and association. With these foundational rights, the public can speak out and organize collectively to influence the government's environmental decisions. (p. 698)

The scholarly and practitioner literature is filled with examples of adequate and inadequate access. From these we can tease out a preliminary list of general grammars of access. At a minimum, it seems that access should be characterized by: an attitude of collaboration, convenient times, convenient places, readily available information and education, diverse opportunities to access information and education, technical assistance to gain a basic grasp of the issues and choices, adequate and widely disseminated notice, early public involvement, and ongoing opportunities for involvement. To this preliminary list can be added other general elements as well as creative approaches to provide voice that accommodates specific contexts.

Standing

This is standing not in the legal sense. It is the civic legitimacy, the respect, the esteem, and the consideration that all stakeholders' perspectives should be given. Access and standing are mutually dependent on each other to achieve influence. Keep in mind, however, that taking the opportunity to submit written or oral comments at a public hearing, for example, is a practice of access. It puts me in a place where I might have standing, but simply having access does not assume that my comments will be given standing. Access is easy to provide. Standing is far trickier. From the existing literature, a preliminary list of general grammars of standing would include: opportunities for dialogue and deliberation; active listening; courtesy, or an absence of discounting verbal and nonverbal behavior; early and ongoing voice; clear parameters of expectations for authority of participation (e.g., how outcomes of participation will be accorded standing in the decision-making process); clear parameters of investment (e.g., how long will a task force be active?); collaborative room arrangements (e g., attention to non-intimidating proxemics); reflection of genuine empathy for the concerns of other perspectives, dialogue, debate, and feedback. Again, to this preliminary list can be added other general elements as well as creative approaches to standing that accommodate specific contexts.

The Interdependence of Access and Standing

If I have access to a process but my participation in it is not accorded standing, then I might as well not have access. The disparity between expectations and experience will make me angry, skeptical, and distrustful. Without the civic legitimacy of standing, I certainly have little expectation of influencing an outcome. Similarly, if I might have standing but I have no access to a space that holds the potential that my standing may be recognized, I have little expectation of influencing an outcome. Denied one or both, I will become frustrated, angry, and increasingly antagonistic and aggressive in creating the space in which I can claim my access or standing. If I have no access or standing, I will find ways to make the others pay attention to my ideas, choices, and preferences. I will become more skeptical of information that is provided to me and will distrust and/or dismiss the sources of the information. Dueling science claims will increase because technical and health risk arguments are allowed, whereas my sense of intrusion, being steamrolled, my vision for my community, or my great attachment to a place is not. As Carpenter and Kennedy's (1986/1988) "spiral of unmanaged conflict" illustrates, it will become more difficult to convince me that you or your information can be trusted. As the distrust and frustration and anger increase, a number of things will occur. The number of issues at stake will increase, the number of participants will increase, the moral righteousness of my position will increase, the investment that I am willing to make will increase, my position will become entrenched, a moral tone will characterize my position, norms of behavior will become relaxed, and opportunities for conciliation and resolution will diminish.

Influence

Influence is the outgrowth of access and standing. By influence, I do not exclusively mean that I was successful at convincing, strong-arming or manipulating others to achieve what I think is the ideal outcome, my position. Influence is not just getting my way, although at times I may. It means that my ideas have been respectfully considered along with those of other stakeholders and my representative or I was part of the process that, for example, determined decision criteria and measured alternatives against it. My idea may or may not be incorporated in whole, but access and standing have allowed an open consideration of what's at stake for everyone as priorities are set and solutions explored. From the existing literature, a preliminary list of general grammars of standing would include meaningful decision space, transparent process that considers all alternatives, opportunities to meaningfully scope alternatives, opportunities to inform the decision criteria, and thoughtful response to stakeholder concerns and ideas.

When we consider the synergistic dynamics of access, standing, and influence, it allows us not only to monitor existing processes but to design better processes as well. It allows a benchmarking system by which to take the pulse of a process and see if it is healthy. Consider the following scenarios, all based on environmental public policy decision-making processes. They are based on real situations that are resolved, ongoing, or in the design phase. Consider what is missing: access, standing, or influence. Project the consequences of continuing in the current mindset and process. Consider what might have been done or could still be done to achieve greater effectiveness by providing for the practices of voice, standing, and influence. What would the tactics, or grammars look like for each situation?

> I am a fishing guide who feels that cormorants are eating all the smallmouth bass. My business is at stake. I have access to public information meetings and I thought I had standing as a member for two years on a task group that provided management recommendations to the state and federal governments who ignored them. I became increasingly angry and frustrated so some buddies and I went out and slaughtered nearly 3000 of those birds. Now maybe I'll have some influence and the government will pay attention to me. It doesn't matter if the enviros have filed a lawsuit; I have a Congressman who's introduced a bill to hunt those birds.
>
> I'm the town supervisor of one of the 41 municipalities in this diverse watershed. Some watershed group has been meeting for three years and now they've issued a watershed management plan. Who do they think they are? They've got another thing coming if they think we're just going to go along with their ideas.
>
> We are the group of municipal officials who have worked over the last three years to develop a watershed management plan. Now we want to implement it. Can't we just form a not-for-profit organization to do this? Why do we need to worry about other groups in the watershed? We'll have the authority to make it happen. The municipal officials who haven't been a part of this will just have to go along. Let's just implement this thing.
>
> I'm a town board member in this municipality where a hazardous-waste company wants a zoning change to allow it to take solid waste as well as hazardous waste when it expands into its approved extra 75 acres. We thought the scoping from the original permit granted years ago was sufficient—no new issues, we thought—so we didn't hold a scoping meeting on this zoning issue. Last night was the public hearing on the draft environmental impact analysis and boy, you'd think we

> wanted to put nuclear waste in their houses. There was standing room only and the citizens were just crazy angry. Now we don't know what to do. The board had already agreed to accept the DEIS (draft environmental impact statement) and we're afraid that we'll be facing a lawsuit if we reject it now this late in the permitting process.

To demonstrate the practical value of the TOV for assessing existing processes, the public hearing, the most common process for involving the public in environmental decision making, is a good candidate. The public hearing provides voice, but is lacking in real or perceived standing or influence.

PUBLIC HEARINGS

By a wide margin, the most common form of public participation, from the local to the national level, continues to be the public hearing. In 1984, Cole and Caputo noted that it was "among the most utilized forms of citizen participation" (p. 415). In 2003, the generic public hearing is still assumed to be an appropriate and effective venue for public participation. It is a familiar structure and the logistics are manageable to ensure maximum control along with maximum detachment from the public, from the beginning to the end of the permitting process.

As do all state versions of the federal National Environmental Policy Act of 1970 (NEPA), New York State's 1976 State Quality Review Act (SEQRA) provides for the option of a public hearing in addition to the required 30-day public written comment period on a draft environmental impact statement (NYS Environmental Conservation Law 8-0109). This option is exercised at the discretion of the designated lead agency in the action. The formal public hearing is named as the venue in which citizens can verbally express their support, opposition, or concerns regarding the DEIS. Although it sometimes happens, officials are obligated to answer these concerns in the final environmental impact statement (EIS) before the permit is issued, but are under no legal obligation to incorporate any of the suggestions or desires into the final EIS nor to reject a project application together if citizen opposition is strong.

Before the mandated written comment period and the optional public hearing on a DEIS, permit applicants (e.g., developers) and municipal officials have typically met several times. Developers have had the advantage of several exclusive audiences with board members to promote and modify their proposal. Board members, in turn, have had access to the developer's scientific and technical support. In addition, they often have the benefit of municipal staff to consider utility, traffic, and other infrastructure requirements and consequences.

At these critical, early stages, the public is not a player unless citizens or stakeholder representatives are individually invited to be part of an optional scoping process. The scoping option in the SEQRA process allows officials to convene a meeting of invited stakeholders to help determine the issues of environmental concern to be addressed in the DEIS.

With or without a scoping session, the developer and the town begin investing human and financial resources into compiling the scientific, technical, architectural, and economic data. However, this process is not structured to assess other less tangible consequences that proposed changes to the land might have on members of the community, such as aesthetics, sense of place, community identity and vision, or a sense of intrusion and powerlessness.

The completed DEIS is then ready for the mandated written comment period and the optional public hearing. The DEIS, often several inches thick or even several volumes long, is put on reserve at the nearest central library, the State Department of Environmental Conservation offices, and the local municipal office to be examined during business hours. The required notice of the comment period and public hearing is placed in the Environmental Notice Bulletin, a weekly publication put out by the New York State Department of Environmental Conservation and available at central libraries or by annual subscription or email. Sometimes town boards are locally required to place a notice in the legal classified section of the local paper. Sometimes they do this as a public service. Larger municipalities may issue the announcement to media outlets as well, but no obligation compels newspapers to showcase it.

Proxemics of the Public Hearing

The proxemics of the typical local public hearing are theater style. Municipal officials sit at the front of the room, often on a raised dais facing the audience sitting in rows. In the front row are typically the representatives of the permit applicant (e.g., engineers, and architects). Citizens sit behind them. Communication is usually one-way. First is the formal presentation of the development plan and its architectural renderings to the audience. Then, if anyone wishes to testify, he or she waits to be called according to the order that he or she was assigned on arrival. Sometimes a microphone is provided, and testimony is given as directed from the center or to the side of the room, or from a standing position at their places. Sometimes testimonies are limited, for example, to three minutes. Occasionally, an official or developer representative may respond to or ask a question, but dialogue or debate is neither routine nor expected. On controversial issues, boos, hisses, cheers, and the display of signs and costumes are not uncommon. In the end, citizens are thanked and the process ends. Sometimes the board takes a vote to accept the DEIS immediately and then adjourns.

Arnstein (1969) early characterized the public hearing as a weak form of public involvement in her ladder of citizen participation. Cole and Caputo (1984) lament the existence of only a "very few serious and systematic attempts to evaluate the effectiveness of citizen participation in general, or of public hearings in particular" (p. 405). Regarding public hearings in particular, Checkoway (1981) notes that there are "few studies which address the effectiveness [of the public hearing] as a participation method" (p. 567) even though there are estimated to be tens of thousands of this most traditional public participation method each year. "Yet evaluations of the comprehensiveness and quality of participation receive a relatively modest amount of attention from social scientists. The social issues addressed are fundamental; the extent of the research available disappointing" (Crotty, 1991, p. x).

Whether the stories come from policymakers, legislators, local town planning boards, citizens, or agency personnel, anecdotal evidence abounds that frames, for example, the public hearing process, in skeptical to scathing terms. It is becoming increasingly clear that the public hearing process, for example, often acts as a trigger of citizen frustration, anger, and even conflict.

Checkoway (1981) is explosive about the manipulative effects of public hearings. Fiorino (1990) asserts that they were effective in legitimating agency decisions that had already been made. Heberlein (1976) notes the intimidating proxemics for those unaccustomed to public speaking. Hance, Chess, and Sandman (1995) note that they may be favorable to interest groups grandstanding rather than commenting. Mazmanian & Nienaber (1979) note the reactive position that public hearings force citizens to assume. Cole's and Caputo's (1984) longitudinal study of eight years of public hearings concerning the General Revenue Sharing Program found that "the public hearing may permit sanctioned isolation of agencies and agency officials seeking as little program and policy change as possible" (p. 415).

Most citizens who attend local public hearings do so to represent a particular perspective on an issue, yet the public hearing format is not conducive to dialogue, discussion, learning, and creative problem solving. Consequently, the public hearing is often marked by tension and often serves as a common flashpoint for conflict escalation, so much so that dispute resolution expert Susan Carpenter (1996) calls it "one of the most destructive and deadly processes ever devised."

Observers of this process often conclude, and I agree after observing nearly a hundred public hearings myself, that "despite hundreds of federal regulations for public involvement programs and the expenditure of millions of dollars, the most common participation procedures, such as public hearings and advisory committees, are more often the most wasteful and useless" (DeSario & Langton, 1987a, p. 14). Bob Wieboldt (1995), former director of the influential New York State Home Builders Association, characterizes state and local decision making and policy processes, especially the public hearing, as "fumbling in the dark and stupid. It's the worst way to get at the truth." According to one state agency official who often deals with public hearings, the process "is a complete sham. Everyone knows it's just a rubber stamp process on a decision that's already been determined." Another says that "officials don't really want citizens to be part of the process. It's a ritual thing to make them think that they have a voice, but [the proposal] is almost always a done deal." And the citizens know it, too.

Although a long random list could be compiled of the reasons public hearings on contentious issues are ineffective, some would be case specific and some would be vague. The TOV allows us to assess whether this or any public involvement process is hitting on all cylinders, so to speak, and what needs to be tuned up. Perhaps an entire engine replacement is indicated. By identifying and clustering the grammars of the ineffectiveness of the practices of access, standing, and influence afforded by the public hearing, we can better understand and assess why it is such a lightning rod for stakeholder discontent and make a stronger argument for improved practice or more collaborative processes.

Applying the TOV, one could ask the degree to which stakeholders had access. Yes, they had access to a process site and they had the opportunity to participate individually. They could submit written comments or register their opinion in person. They were most likely safeguarded as they spoke. Their words were recorded, transcribed, and would be entered into the historic written and perhaps video record of this event. They had access to a space where the potential existed to have standing. Did they have standing, that is, civic legitimacy? To what degree were their words honored, that is, given due consideration and reflection, really listened to? If we accept the anecdotal as well as the empirical evidence, very few participants, whether administrators, developers, or citizens, feel that anybody was really listening at most public hearings. Citizens often comment that that's why they act dramatic, loud, obnoxious, emotional, and even threatening—they know that no one is really listening and they are trying to shake them into paying attention, to make them realize that they are serious about these issues. They are creating their own standing by creating media events, bolstering their organization, appealing to other citizens to join them, and trying to intimidate officials into thinking of further repercussions.

The placement of the public hearing at the end of the SEQRA process, the time limit of the individual comments, the arrangement of the room, the reactions of officials, the inaccessibility to the thick DEIS, the lack of technical support to help citizens understand—all of these reinforce the notion that citizens do not have much standing in the process, even if they have limited access to it. Most municipal officials are paid little if anything for their office, often hold other full-time jobs, and are provided few if any opportunities to gain confidence in other forms of participation (Church, 1995). No wonder they default to the traditional if unsatisfactory forms that make them reluctant to engage in unfamiliar forms to interact with these citizens who appear to be hysterical, irrational, and ignorant. In this self-defeating cycle, citizens have very little civic standing; hence, their access is diminished as well.

To what degree do they have influence? With few exceptions, not much if any. The decision space is small because most if not all of the decision has already been made. Developers might tweak a parking area, plant more mature trees, take additional care buffering a wetland during construction, or change some minor specifications, but they typically do not withdraw an application. If the lead agency denies the permit at this point, it is possibly laying itself open for litigation by the permit applicant. By the effectiveness measure of how much the citizens were able to affect the criteria by which decisions were made to change the final EIS, the citizens had very little influence. How could they? They had no standing.

CONCLUSION AND DIRECTIONS FOR FUTURE RESEARCH

Environmental laws, regulations, executive orders, and treaties explicitly call for effective and meaningful public participation in environmental decision making. The fundamental measure of effective and meaningful public participation processes is the degree to which the process supports trust. However, trust is difficult to generally operationalize across individual situations and contexts. Hence, case studies and training materials offer disconnected lists of what characterizes effective and ineffective processes. A general, practical theory is needed to provide a common template to building, enhancing, and remediating trust in public participation processes concerning environmental issues.

The TOV is offered as a general theory for inquiry into the grammars (e.g., the specific tactics, behaviors) of the effective public participation practices of trust, that is, access, standing, and influence. The TOV offers a useful mechanism by which to design and assess the effectiveness of public participation processes as well as to analyze an escalated conflict to identify the raw spots and structure a process for resolution. It allows for learning from the opportunities and unique contexts of individual cases while pulling common threads through a number of cases. Rather than identifying piecemeal elements of individual accounts of processes, the TOV suggests that these be thought of as the grammars that characterize the more general practices of effective public participation—access, standing, and influence.

I invite others to play with this general theory to see how far it can stretch. For example, what are other grammars of effective and ineffective public processes? To what degree does the TOV help in understanding the effectiveness or ineffectiveness? How can the TOV be useful in guiding process design? Are there predictable (in a general sense) outcomes when different combinations of the elements are present or missing or if one is severely out of proportion to the others? Is it as useful for siting issues and hazardous waste cleanup situations as it is for local land use decisions? Is it equally useful at different levels of participation such as local, regional, national, and international? How can it be used to support the assessment of escalated conflicts to determine the fruitfulness of convening a consensus-based process to resolve a conflict?

In 1979, the Advisory Commission on Intergovernmental Relations concluded from its survey of American public participation objectives that one of the ultimate goals of the increased expectations for public participation should be "to change governmental behavior so that governmental units respond better to citizens' needs and desires and refrain from the arbitrary, capricious, insensitive, or oppressive exercise of power" (p. 61). Cole and Caputo (1984) contend that this should be pursued by "forging at all levels a more accountable, more responsive, more democratic government, especially in administrative and bureaucratic activities" (p. 44). Accessibility of the process and the responsiveness of it to the needs of those affected, not the efficiency or rationality, should be the measure of the decision (Kweit & Kweit, 1984). Like many others, Kemmis (1995) calls for new civic forms to rekindle the connection of citizens to communities and Schwerin (1995) calls for a renewed sense of citizen empowerment. The challenge is having a general guide through the individual contexts.

As Cronen (2000) notes, "Instead of thinking theory as prior to action, think of it as an instrument within the course of action. Rather than thinking of inquiry into a situation as a way to support a theory, think of theory as a way to improve a situation" (p. 23). The Trinity of Voice fosters that link between theory and practice and offers the opportunity to build arguments to foster effective process.

NOTE

1. A plethora of articles, manuals, and training guides attest to the many negative consequences (for policy as well as for relationships) of performing token or placating public participation or neglecting it all together. There are fewer writings that offer case studies of effective process, but these usually offer either (1) tactics that were specific to the case (e.g., weekly radio shows), (2) tactics that were generic (e.g., open information flow), or (3) conceptual (e.g., trust building). See the following Web sites for extensive bibliographies and Web links: http://www.pin.org (Public Involvement Network), http://www.iap2.org (International Association for Public Participation), http://www.ncl.org (National Civic League).

CHAPTER 2

A SOCIAL COMMUNICATION PERSPECTIVE TOWARD PUBLIC PARTICIPATION: THE CASE OF THE CISPUS ADAPTIVE MANAGEMENT AREA

AMANDA C. GRAHAM

In a democracy, the process of making decisions that affect public policy and the administration of public goods and services is expected to involve the public. In the latter half of this century, a body of legislation has taken shape that requires and structures citizen participation in policymaking.[1] Regulations governing policy implementation in fields ranging from health care to transportation to the environment are incomplete without the inclusion of guidelines for some form of public participation.

This trend toward formally incorporating citizen voices in decisions that affect policymaking and management activities for the public welfare has been clearly visible in natural-resource management for over a quarter of a century (Wellman & Tipple, 1990). The year 1970 saw both the creation of the Environmental Protection Agency (EPA), the first federal agency overseeing national environmental resources, and the enactment of the first sweeping mandate for the involvement of the public in the decision-making processes of federal agencies responsible for the administration of natural resources, in the form of the National Environmental Policy Act (NEPA) (Twight, 1977). NEPA requirements continue to provide the broad framework within which federal administrative agencies implement public participation programs. Since NEPA, legislation and regulations to govern the management of natural resources have proliferated, including such landmark legislation as the Clean Air Act, the Clean Water Act, the Endangered Species Act, the Forest and Rangeland Renewable Resources Planning Act (RPA) and the National Forest Management Act. As federal agencies such as the U.S. Bureau of Land Management (BLM), the U.S. Forest Service (USFS) and the U.S. Fish and Wildlife Service implement legislation that applies to their activities, they are required to follow both NEPA guidelines for public participation and regulations and guidelines internal to their own agency.

Almost thirty years of agency interpretation and application of these guidelines have produced numerous strategies and techniques for involving the public.[2] Yet lawsuits against agency actions, filed by citizens or environmental groups, have been an increasingly visible feature of agency–public interaction, virtually since the enactment of NEPA itself (Gericke & Sullivan, 1994; Kessler, 1991; Magill, 1988). One widely acknowledged example of gridlock in forest management in the Pacific Northwest received close nationwide scrutiny and precipitated presidential involvement in 1993.[3] These events vividly indicate the intensity of citizen concern about public participation and the power found in filing lawsuits against agency work. To some observers, legal appeals against agency action have come to be seen as an unfortunate but unavoidable part of doing the business of natural-resource management. To others, the frequency and sheer number of lawsuits, as well as their ability to bring agency activity to a grinding halt, suggest that natural resource "professionals are not responding to public wants," and official forums for public participation processes are not working effectively (Magill, 1988, p. 295). At least partly in response to this phenomenon, public participation processes are the focus of considerable attention among agency professionals, citizens, environmental groups and scholars (Blahna & Yonts-Shepard, 1989; Gericke, Sullivan, & Wellman, 1992).

From the vantage point of communication research, two problems are visible in approaches to public participation. One problem is that some public participation processes are founded, at least in part, on a narrow perspective of communication. A growing trend in contemporary communication emphasizes the dialogic (Baxter & Montgomery, 1996), social (Leeds-Hurwitz, 1995), co-constructed (Shotter, 1993; Stewart, 1995b) nature of the process of communicating. In part, federal resource management agencies are responsive to these features. Yet at the same time, they often appear to treat communication as the transmission and exchange of static, preformed information and opinions. A brief analysis of excerpts from a 1992 *Public Participation Handbook* (USDA-FS, 1992) for employees of the Forest Service and a 1991 (USDA-FS, 1991) report on internal and external Forest Service communication offers a window on these contradictions within the agency's conceptualization of communication.

In this manual intended to provide guidance for agency personnel who "plan and carry out public participation programs and actions" (USDA-FS, 1992, p. I-1), agency authors emphasize the "continuous" (p. I-3) and "responsive" (p. I-1) nature of public participation and include *"public understanding about and participation in* the planning and decision process by providing information" as a primary agency goal (p. I-1; emphasis added). Managers are encouraged to "make public involvement an integral part of any planning, program, or project and not a separate feature," to "begin public participation at the earliest possible stages," and to remember that "it is everybody's job" (p. I-1; p. I-3). The authors also state that at the most "basic level, public participation is nothing more than two-way communication. These communications occur every time we answer the telephone, respond to a letter, or meet with a member of the public" (p. I-3). This two-way nature of communication is noted in several places, including in the following excerpt in which agency authors introduce a section on interactive public involvement techniques with the observation that

> In the last two chapters we discussed techniques that were basically one directional. They either gave information to the public or collected it. Here are some examples of techniques that supply a give and take in the information flow. (USDA-FS, 1992, p. I-67)

An agency report titled "Improving Communications and Working Relationships," which predates the 1992 edition of the *Public Participation Handbook*, identifies another element of an alternative view of communication "many Forest Service people and folks outside the Service feel an urgent need *to build better relationships* and to improve our *communication*" (USDA-FS, 1991, p. 2; emphasis in original).

Several open and dynamic aspects of the USFS approach to communication are apparent here. First, there is a clear attempt to define communication as two-directional. The agency is clearly interested in bringing information in as well as sending it out. Two-way communication is intended to encourage communication from the public into the agency. Second, these passages also suggest the ubiquitous, ongoing aspects of communication: it takes place in many shapes and forms, it is a continuous process, and everyone is engaged in it. Finally, the excerpt from "Improving Communications" links communication with building relationships.

This analysis highlights three underlying assumptions about communication in this selection of Forest Service documents: expanding beyond one-way communication, communication as ever present and continuous, and communication as connected to relationship building.

Yet internal contradictions are also apparent. The first passage quoted here suggests that the route to developing understanding among the public is "by providing information" to them. The last passage echoes this basic approach by referring to a "give and take in the information flow." Providing, giving, and taking information suggests that the information the Forest Service seeks to give and take exists prior and external to the forum in which they are engaged with the public. These comments reflect an assumption that information is a discrete, identifiable, transferable object. Although it is certainly true that the people bring experiences and ideas with them to the table, this approach—treating interactive events as information exchange—is limited by its failure to acknowledge the power of communicative interaction to shift, change, and create shared knowledge and values critical to the joint development of a course of action. An underlying understanding of communication as a means to exchange or collect static pieces of information appears to be a significant aspect of the agency's approach to public participation.

Later passages in the *Public Participation Handbook* (1992) treat it primarily as an opportunity to gather data for later analysis by agency staff and internal evaluation and incorporation into decisions.

> Once public comments have been collected, they should be analyzed and summarized, and the decisionmaker should evaluate the findings. The object here is to zero in on specific information that later might be used in developing alternatives. (p. I-11)

> The analysis of public response involves the consolidation of what the public had to say and its presentation in an abbreviated format for evaluation by the decisionmaker. (p. I-14)

In addition to reemphasizing the notion of communication as information based, these passages suggest a fragmented, rather than holistic, approach to "what the public had to say." "Zeroing in" on specific information creates an opportunity for an agency analyst or decision maker to disregard the rich context within which citizen ideas and experiences are located. The insertion of an agency or professional analyst into the public participation process reinforces the notion that the point of the process is to solicit already formed "information" from citizens, and can serve to further distance decision makers from the complexities of the deliberative process.

In a comprehensive analysis of the evolution of public participation in the USFS, Shannon (1991) observed that "for the most part the techniques of involvement and participation . . . remain primarily techniques for getting or sending information" (p. 25). Shannon critiques the agency's autonomous decision-making process as hierarchical and therefore "based on the downward flow of information" (1991, p. 26).[4] Taken together with the analysis preceding, Shannon's comments suggest that public participation often reflects an understanding of communication as a process of collecting preexisting knowledge and values about natural-resource management issues and direction.

A second problem that appears from a communication analysis of participation is an apparent gap between agency rhetoric about public involvement and public participation, and the manner in which efforts to involve the public actually take place. Shannon (1991) speaks to this disconnect when she observes that "addressing the confusion between an open, deliberative process and the jurisdictional responsibilities of the agency needs to occur. . . . The roles of the public and the roles of the agency in a democratized administrative structure have never been clearly articulated" (p. 85). Similarly, Magill (1988) asserts that agency personnel have a tendency to listen without hearing, hesitate to cultivate personal contact with citizens, and are unable to develop long-term relationships with citizens due to frequent transfers from region to region. He concludes that many resource managers "remain insensitive" to the public as a result of these (and other) factors, despite increased emphasis and occurrence of public involvement programs (p. 295). He terms this "communication failure" (p. 295) and suggests that it causes alienation among citizens from agencies and agency efforts to involve the public. Shannon (1991) and Magill (1988, 1991) both point to a lack of social science (as contrasted with natural science) and communication training among forestry professionals as one root of these difficulties in agency–public communication.

Reporting on the public participation component of the broad-based social assessment conducted by the Forest Ecosystem Management Team (FEMAT) in 1993,[5] Clark, Stankey, & Shannon (1999) note that during critical stages of forest planning, typically "there was little or no systematic contact with the public most directly affected by those decisions" (p. 252). Clark et al. conclude their section on public participation with the following interpretation of the basic problem facing formal agency–citizen interaction:

> Although an array of legislative requirements exists for acquiring public involvement in resource planning and management, well-established programs and policies that effectively integrate public input into decisionmaking remain elusive. Despite these legislative mandates, and the accompanying agency efforts to meet legal requirements, there is substantial evidence that the goals underlying public involvement programs—informing people, soliciting their ideas, integrating their concerns into decisions, in short, being responsive to those who are the owners of the public lands—are not met in practice. (p. 253)

What these three sets of observations about the USFS make clear is that although the agency has articulated and accepted an inclusive, two-way, continuous, and responsive approach to working with the public, the reality of implementing has fallen short of achieving these qualities.

Public participation processes in natural resource management provide a conceptually and pragmatically interesting site for social communication research.[6] As this brief review indicates, communication analysis can render visible previously unrecognized themes and inconsistencies that underpin and profoundly influence approaches to public participation. This chapter engages a conceptual approach in assessing the practical problem of how to do public participation and suggests that a richer understanding of communication opens up the possibility for enhanced engagement between the government and the public in the environmental arena. Specifically, this chapter analyzes an empirical example of public participation in environmental management from an understanding of communication as socially constructed, dynamic, and emergent. I believe that this approach can clarify the centrality of interpersonal relationships and the co-constructed nature of understanding and knowledge, and can thereby enhance and empower public participation processes designed to fulfill the expectations of the agencies and the citizens who join to manage and democratically govern natural resources.

SOCIAL COMMUNICATION: SOCIALLY CONSTRUCTED, DYNAMIC, AND EMERGENT

The central tenet of social constructionism is that human understanding of reality is created or constructed through interaction and interpretation (Leeds-Hurwitz, 1995, p. 7). Two conclusions follow from this premise. First, that there is not one single truth, perspective, or understanding of reality, but that people co-construct reality or meaning in their communication. A social constructionist view of communication accepts "the integrity of multiple, valid, and contradictory perspectives engaged in dialogue" (Baxter & Montgomery, 1996, p. 47). Second, the process of communicating therefore becomes a necessary focal point for understanding how people come to understand each other and the worlds around them. Social communication is what happens between and among people. Communication from this perspective is the process of relating: it does not exist except between or among particular people (Baxter & Montgomery, 1996; Shotter, 1993; Stewart, 1995a). Further, one set of communicative dynamics and relations cannot be transferred wholesale from one interacting group of people to another. People as individuals are neither interchangeable nor completely independent ("atomic"); identities themselves are shaped in interaction (Leeds-Hurwitz, 1995, p. 8; Shotter, 1993, p. 45). If reality and identities are created through social interaction (communication), it can also be said that communication is a constructive act (Stewart, 1995a, p. 113). In sum, then, as people engage in communicating with one another, from a social constructionist point of view they are fundamentally, collaboratively engaged in a process of creating, or constituting, interpersonal relationships, individual and collective identities, and social "worlds" (Pearce, 1995, p. 98; Shotter, 1993, p. 36; Stewart, 1995a, p. 111).

The notion of emergence underscores the evolutionary nature of communicatively accomplished meanings. Meanings are not static, repeatable commodities, but are negotiated as interactions unfold. This construct foregrounds how significantly context elements such as gender, ethnicity, family background, physical environment, and relationship history help to shape the ways people talk with one another. However, from the standpoint of social communication, the nature and quality of people's talk are not fully predetermined by context factors. Rather, context is "dynamically sustained" during communicative interaction (Shotter, 1993, p. 2). Thus, the course of conversations—the understandings that are simultaneously developed—and interpersonal relationships are continuously formed and reformed at the convergence of context, the skills and identities of the conversation partners, and the dynamism with which they interact (Baxter & Montgomery, 1996, p. 47). Both relationships and understanding are "partial, tentative, and changing" (Stewart, 1995a, p. 125).

A Social Communication Perspective on Public Participation

This understanding of communication as socially constructed implicates a version of public participation distinctively different from traditional approaches to bringing citizens and government employees together.

The contingent nature of knowledge, the decentralization of responsibility, and the centrality of interpersonal relationships become defining aspects of public participation processes rather than afterthoughts or interesting but inconsequential academic constructs. Together these concepts reconstruct the terms *participation* and *collaboration* as they are often used in the discourse of federal agency–public interaction.

Openness

Public participation has traditionally treated information as a stable, inert commodity produced by experts (Shannon, 1991). Public participation informed by a social communication perspective would be open to, and would value the inclusion of, knowledge, experience, and perspectives from all participants, including general citizens, tribes, environmental groups, industry and others. Participant planners would be as attentive to knowledge and ideas about social aspects of resource planning as they would to ecological and technical knowledge. In maintaining a critical stance toward knowledge gathered and developed during the process of deliberation, participants would seek to understand the context of the knowledge in their ongoing process. They would recognize that even during their discussions, they were creating and shaping knowledge about themselves and their interests and would return to and use this experience throughout their engagement. Moreover, participants would endeavor to remember that their knowledge can never be complete or final. The data they gather would always be open to interpretation and addition.

Shared Responsibility

When one party, such as a federal agency, has primary responsibility for creating and facilitating a process of public participation—and when that party has a distinct and narrow reason to be invested in that process, namely to gather focused input from the public—other parties are then forced to take responsibility only for themselves and their own distinct agenda. A social communication perspective on public participation suggests that all parties involved should work together to develop a sense of shared responsibility for a set of common objectives. This will not take place when Party A asks Parties B and C to join in meeting A's objectives. Rather, there must be a forum in which all parties engage together in creating objectives in which they all experience ownership. Jointly developing and committing to these objectives becomes a necessary component of the process of working together on a particular issue in a version of public participation fully informed by a social communication perspective.

Interpersonal Relationships

Just as interpersonal relationships form the basic unit of analysis in social communication, so do they also serve as the basic unit for a collaborative program of agency–public involvement. The interpersonal level that is the site for building understanding and identities also becomes the place to engage people in dialogue about their objectives, their concerns, and their knowledge for the purpose of democratic deliberation and decision making. The cultivation and nurturing of relationships develop a forum in which participants can respond to each other with an understanding of mutual responsibility and respect. Establishing and valuing interpersonal relationships creates a larger capacity for positive action than does treating people as anonymous, interchangeable strangers. Placing relationship development at the center of the public participation process also creates capacity for ongoing and cumulative joint activity.

An opportunity for empirical examination of a public participation process from a social communication perspective was provided by activities in the Cispus Adaptive Management Area, a localized component of a major experiment in transforming agency–public interactions (Graham, 1996). The author used participant observations, workshop evaluation questionnaires, interviews with agency employees and members of the public, and review of published documents to examine public participation activities in this case and assess the degree to which these activities reflected a socially constructed understanding of communication.

Case Study: Public Participation in the Cispus Adaptive Management Area

In 1994, a portion of the Gifford Pinchot National Forest was designated an Adaptive Management Area (AMA) under Option 9 of President Clinton's Northwest Forest Plan. The Cispus AMA, along with the other nine AMAs located throughout the range of the northern spotted owl in the Pacific Northwest, was charged with actively experimenting with both social and ecological management techniques. More specifically, the ten AMAs were intended to develop and implement more collaborative modes of public participation and citizen involvement in AMA planning and management (USDA-FS & USDI-BLM, 1994). The Cispus AMA responded to this charge with a series of workshops in 1995 in which citizens were engaged in small group discussions with each other and local Forest Service employees. As was the case with all the AMAs, a coordinator was appointed on the Cispus to facilitate planning and management on the AMA and coordinate agency staff and public involvement in Cispus AMA activities.

Over the course of three months and a total of six public workshops, adults of a wide range of ages, experiences and perspectives participated in the AMA planning and public participation process during the spring and summer of 1995. Participants came from unincorporated communities adjacent to the AMA as well as from metropolitan areas up to a two-hour drive away. A total of approximately 80 individuals attended at least one of the six workshops, with a core group of approximately 20 individuals attending nearly every meeting. All workshops were held in a local high school library. In addition to the workshops, the Cispus AMA periodically published and distributed a newsletter on AMA activities to a broad mailing list including every workshop participant as well as citizens and groups who expressed interest to AMA.

To engage the public, the U.S. Forest Service incorporated four basic elements into the workshops: individual completion of open-ended worksheet questions, small group discussion, sharing highlights from small group discussion with the large group, and occasional presentations by U.S. Forest Service staff. These and other workshop elements reflected the local Forest Service's effort to use a collaborative learning approach to the Cispus AMA planning process.[7] A local U.S. Forest Service planner facilitated most of the workshops and referred at each one to a written set of workshop guidelines to help maintain a productive working atmosphere among the group. Workshop attendees seated themselves at four-to-six-person tables. Numerous U.S. Forest Service staff attended the workshops as participants, some out of job-related interest, and some out of personal interest. Besides the facilitator, all agency staff participated in the workshops alongside members of the public. Local agency leaders ("district rangers") participated in every workshop.

To explicitly address the degree to which public participation on the Cispus AMA reflected the qualities of social communication, the remainder of the case discussion in this chapter is organized by the three critical features of public participation from a social constructionist viewpoint previously discussed: openness to diverse participation and perspectives, shared responsibility and investment, and prioritization of interpersonal relationships.

OPENNESS

From the beginning, the coordinator of the Cispus AMA persistently foregrounded the social and human component of planning and management as well as the value of the perspectives, ideas, and knowledge of nonagency individuals and groups. At the first public workshop on April 1, 1995, the coordinator identified the purpose of the network of ten AMAs as a method for communication, information sharing, and the development of social relationships and noted that "today is the beginning of that process, and we'll need to keep exploring, and finding out ways we can best work together" (notes 4/1/95). At this time, the coordinator also emphasized that "the most important product from the AMA is not timber. It is shared learning, so that we can communicate and understand the consequences of management on all the system, including the social [component]" (notes 4/1/95).

Local U.S. Forest Service leaders also embraced the knowledge brought by citizens to the workshop series. One outside agency observer noted that the very presence of the two district rangers at nearly every workshop—plus their active, personal involvement as workshop participants—were in themselves indicators of a strong local agency commitment to the public's perspective (personal communication, July 1995).

The first issue of the AMA's newsletter, the *Cispus AMA Quarterly*, placed the public and public participation in a central position in AMA activities. The following excerpt from this newsletter suggests an open and inclusive stance on the part of the local Forest Service toward the public:

> The President's Forest Plan emphasizes public involvement as being critical to the planning process. To us, it just makes good sense. For many of you, a strategy developed in this process may affect what happens to the forest in your backyard. It might affect how you make your living, or what opportunities your children or grandchildren will have years from now. That's why your involvement is important to identifying desired future conditions. . . . Collectively we should be able to make the right choices. The AMA is a new, and for us, exciting process . . . a new chance to work together. (Cispus AMA, 1994, p. 1)

This passage also suggests that the agency is open to collaboration and collective decision making with the public, and expresses agency appreciation for the many ways in which citizens' lives are affected by the forest and the way it is managed. The friendly, respectful, enthusiastic tone of this passage implies agency interest in multidirectional sharing of knowledge, and interest in engaged relationships with the public.

A third example of openness is provided by the use of *situation mapping*, a key *collaborative learning* tool,[8] during this process. Situation mapping is a method of visually representing issues, players, activities, and concerns in a diagram showing the relationships and connections between and among them. Key issues and players such as "project planning," "water quality," "local Forest Service," and "local communities" are identified, located in writing on an easily viewed surface such as an overhead transparency, and circled or boxed. Action words denoting the relationships among these issues label lines or arrows connecting the illustrated issues, players, activities, and concerns. Collaborative learning treats situation mapping as a learning exercise rather than as a detailed or comprehensive systems model, although it is intended to convey the basic interrelatedness and complexity that define a systems-based understanding.

Situation mapping was used both during a prepublic involvement training for Forest Service staff as well as during the first workshop with the public. By explicitly incorporating systems thinking into the process of deliberation, the use of situation mapping framed the public participation forum as an opportunity to develop an integrated understanding of critical AMA planning issues that would be responsive to the concerns, experiences and knowledge of citizens.

Finally, many members of the public who participated in the collaborative learning workshops were pleased with their experience. Remarks about how different this forum was from other Forest Service meetings they had attended in the past included appreciation for openness both on the part of the local Forest Service and on the part of fellow workshop attendees. One participant was confident that "this format was obviously better. It proves that there can be public involvement with local input addressed" (interview by author, July 29, 1995). In the words of another attendee, "it was good to be able to talk to other user groups. To share common interests, concerns, and challanges [*sic*]. To hear others drop their prejudices and really listen was great" (April 1, 1995, workshop evaluation). In addition, most participants believed that their comments were important. "Everyone at the meeting seemed to have the sense that their opinion counted, and was being heard. The Forest Service is really trying to see that people are involved," said one participant (interview by author, July 30, 1995).

Shared Responsibility

Although these elements of the public participation process reflected an openness to diverse perspectives and knowledge of citizens, several factors indicate that the process took place within an environment where one party—the U.S. Forest Service—took primary responsibility for shaping deliberations. This facet of the public participation process limited the level of ownership of nonagency participants for the process and its outcomes. As noted in the following section about Interpersonal Relationships, lack of mutually developed investment could also potentially jeopardize relationships built during the collaborative learning process.

The critical issue of responsibility is shaped by the federal document that guides the implementation of AMAs: the Record of Decision (ROD) and its accompanying Standards and Guidelines (USDA-FS/USDI-BLM, 1994). Additional insights into a vision of how public participation in federal natural-resource planning and management should be transformed can be found in the Forest Ecosystem Management Team report (FEMAT, 1993).

The FEMAT discussion of public participation calls for a redefinition of the relationship between agencies and the public to one of partnership instead of hierarchy (FEMAT, 1993, p. VII-114). This partnership entails a fundamental reformation of the way agencies do planning, toward processes "founded on collaboration, powersharing, and mutual learning" (p. VII-115). FEMAT specifically recommends that one of the factors used to evaluate public involvement be the degree to which it is "adequate, meaningful, and participatory" (p. VIII-19).

The discussion of public involvement in the ROD is limited to the following passage:

> AMAs offer the opportunity for *creative, voluntary participation* in forest management activities by willing participants. . . . Our approach to implementing this initiative will recognize and reflect [the differences among them] as we seek to encourage and support the *broadest possible participation.* (USDA-FS/USDI-BLM, 1994, p. 28; emphasis added)

The Standards and Guidelines document that accompanies the ROD and gives specific direction to the responsible agencies emphasizes that communities should be involved early on and continually in the planning and implementation process. It is further noted in the Standards and Guidelines that a specific question to help guide the social component of AMA assessment and planning is "what collaborative process will work best for the *communities of interest to effectively participate in managing* the Adaptive Management Area" (USDA-FS/USDI-BLM, 1994, p. D-6; emphasis added). Although the discourse in the FEMAT document is suggestive of an agency–public relationship characterized by increased sharing of responsibility, the language in the ROD and Standards and Guidelines is more conservative and is oriented toward facilitating the public's involvement in agency-driven processes.

Two key passages indicate that the conceptualization of the roles of groups and individuals outside the agency does not reflect a consistent understanding of *participatory*. The Standards and Guidelines document states that "identification of the desired future conditions *may be* developed in collaboration with communities, *depending on the area*" (USDA-FS/USDI-BLM, 1994, p. D-7; emphasis added). Collaboration and multidirectional engagement are not optional in a social communication framework. More critically, the implementation portion of the Standards and Guidelines specifically states that "while public interest and participation will differ with the issue being considered, *the authority to manage the public lands and resources remains by law with the land management agencies*" (USDA-FS/USDI-BLM, 1994, p. E-15; emphasis added). When contrasted with the passage from FEMAT, a fundamental contradiction is revealed: AMA management cannot be both collaborative and under one authority.

To summarize this analysis, federal documents guiding the implementation of AMAs encourage early, continual, and collaborative involvement of the public. The language of collaboration through creative, participatory processes is suggestive of a dialogic conception of public participation. However, the use of this language to refer to public involvement is diffused throughout the documents, is not consistent in its support of this perspective, and, most important, does not address the fundamental question of how the collaborative approach to public involvement that is suggested will shift the basic differential in power and responsibility. Although it emphasizes a collaborative, engaged, dialogic process, the language in these documents equivocates on the crucial point of whether or not such collaboration involves shared control.

Yet it is important to remember that a central tenet of a social communication perspective is that collaboration itself is not optional. Whether the agency and the public are working together on a level playing field toward mutually satisfactory goals, or, as is customarily the case, they are working together to identify those goals in an atmosphere characterized by the pervasive power differential previously discussed, in both cases they are co-constructing the situation. This collaborative feature of a social communication perspective extends even to conflict and disharmony: "it takes two (or more) to disagree, and thus conflict, too, is necessarily relational, mutual, and in this sense collaborative" (Stewart, 1995a, p. 111).

Public participation literature in the field of natural resources acknowledges this conflict. Yaffee (1996) captures the discomfort in his observation that "an explicit focus on decision-making assumptions, while necessary, opens up agency decision making in ways that are threatening to agency control and experts' sense of professionalism" (p. 725). Although Yaffee advocates the development of nontraditional and collaborative decision-making approaches and organizational capacity for innovation, he concludes that "the final role for government is to *not get in the way*" (p. 726, emphasis in original). These comments echo FEMAT's call for a "fundamental reformation" in federal agency planning and policy-making processes to accommodate genuine collaboration. Shannon's (1991) analysis of the bureaucratic nature of federal planning activities and Kruger's (1996) critique of resource management based on a public philosophy of competitive pluralism[9] also conclude that these structures must be transformed if they are to become consistent with the tradition of democratic deliberation.

Cispus AMA staff members were well aware of the legislative and administrative context for AMA planning and management activities, and they made a point to share this background with the public. At the April 1 workshop, one of the district rangers described the legal "sideboards" that defined the Forest Service's decision space regarding the AMA, and which would act as constraints on the AMA public participation and planning process. Various participants, both public and Forest Service, referred back to these sideboards at several junctures over the course of the spring and summer meetings. Through the district ranger's initial comments and the subsequent discussions about parameters, participants were made aware that their discussions were located within a set of constraints that had some flexibility, but that were outlined by the agency and ultimately up to the agency to arbitrate.

Three additional features of the process of discussion reinforced the separation of responsibilities between the agency and the public. Although the situation mapping exercise served to gather multiple perspectives regarding the future vision of the AMA, the way in which this technique was utilized limited its potential for developing shared objectives and responsibility. At the April 1 workshop, a basic situation map was created prior to the meeting, and public attendees added to and subtracted from the situation map as given. But because control over the original identification of categories and important information was vested in the workshop facilitators, members of the public did not have an opportunity to develop a sense of collective purpose and ownership together with the agency to drive the process from the start.

Each workshop centered on a broad theme, such as "Learning Opportunities" or "Timber and Forest Products." These themes were identified by Cispus AMA staff members. In addition, agency personnel developed a handful of open-ended questions for the public to consider. These questions formed the basis of the worksheets that were completed by each participant, and that then formed the basis for small group discussions. In this format, the public was playing a reactive role as they responded to the agency's questions and areas of interest. This is in contrast to the creative role citizens would be called to play in the design of a process fully informed by a social perspective on communication. Clearly, there was room for creativity at these workshops; individually and collectively, members of the public were encouraged to offer all their ideas, no matter how many or how outlandish. Yet creativity within a preestablished framework is fundamentally different from creative, shared engagement in the development of the framework itself. In these workshops, the framework was provided by the agency.

At the close of each small group discussion period, each individual or group (it varied from workshop to workshop) was asked to share key responses or ideas from their work together with the group as a whole. This large group component of the workshops served a narrow function of reporting as the facilitator clarified and then recorded each comment without discussion among the large group. Especially in light of the potentially insular nature of the individual small group (see "Interpersonal Relationships" later in the chapter), the lack of some sort of discussion or deliberation among the broader group represents a lost opportunity for critical learning and integration among workshop participants. Instead, this aspect of the workshops functioned as a way for the agency to gather information from the public and maintain control over how it would be used. The dynamic of reporting small group ideas and conclusions to the Forest Service for recording echoed the traditionally hierarchical relationship between agency and the public, and reinforced the perception that the agency would be responsible for, and in control of, both the process with the public and the process of putting the suggestions and understandings developed in the small group conversations into place.

Whereas participants largely enjoyed the meetings and assessed them as markedly better than previous public involvement strategies, many citizens expressed a wait and see attitude regarding the long term, in terms of both the form of continued public involvement and whether or not this process would make a difference in forest management. "We'll know five or ten years down the line" whether or not this process developed a successful plan, commented one participant (interview by author, July 30, 1995). Another interviewee said that "at this point, the process feels really incomplete" (interview by author, July 29, 1995). A comment on a workshop evaluation early on in the process sums up a common sentiment among participants from the public: "I was heard and my concerns were addressed, but I have reservations about the USFS ability to answer them on a local level. Your regulations are too strict, and normally do not let local managers resolve issues without interferance [*sic*] from D.C. or the court system" (April 1, 1995 workshop evaluation). These comments reinforced how separate the public and the U.S. Forest Service remained, even after committed participation in six intensive workshops.

Interpersonal Relationships

The opening component of the April 1 workshop, a slide show of the landscape and activities on the Cispus AMA, provided common ground for interpersonal interaction. A local U.S. Forest Service planner took the group on a photographic tour of the AMA, highlighting both the scenic beauty of the area and "the most important aspect of the Cispus AMA—people" (notes 4/1/95). Throughout the slide show, members of the audience frequently murmured comments to each other as they guessed which lake was being shown, or realized they had never seen that particular view (or that particular species) from that location. This workshop component grounded participants in something they all shared: common concern and caring for the land and the life and activities it supports.

A written set of "Workshop Guidelines" served as a sort of bridge between why the group was working together and how they should work together. At each workshop, the facilitator reviewed one or two of the guidelines—such as "listen openly and actively" and "give everyone equal opportunity to speak"—and emphasized that everyone in attendance was there to learn from each other, and thus they should listen to each other and treat each other with respect (notes, 4/1/95). Although few workshop attendees specifically mentioned the guidelines themselves as a helpful component of the workshops, many attendees noted aspects of the meetings that reflected the values and process concerns listed in the guidelines, such as all participants having an opportunity to speak, and treating each other with civility and respect. These qualities characterized interaction throughout the workshop series.

All of the meetings took place in a local high school library. The library was an interesting meeting environment because it combined aspects of a large room (many people could gather at once and both see and hear a centrally located speaker) with aspects of a smaller, more intimate setting (book stacks divided the large room into three or four sections, with one or two tables located in each subarea). The setting contributed an informal atmosphere to the workshops that helped set people at ease, fostered interaction, and focused attention on the small group dynamics.

The nature of the participation of Forest Service employees was highlighted by numerous participants, both Forest Service and members of the public. Several participants noted that the informal approach agency personnel took to interact with citizens in the workshops helped agency staff and the citizens develop a mutual sense of respect and appreciation for each other as individuals. Forest Service employees attended out of uniform, which was noted with appreciation by several members of the public. One participant observed that this helped to make Forest Service employees more approachable. Without preplanning, most tables ended up with at least one local Forest Service employee who participated in the discussion, switching hats between his or her roles of individual citizen and professional scientist or specialist. Forest Service employees participated in face-to-face discussion of concerns and interests. This side-by-side interaction over the course of several meetings enabled the Forest Service and members of the public to establish personal connections that were noted as personally and professionally rewarding. Several participants expressed hope that these connections would continue to be a positive force.

The small group discussions themselves took place after participants completed a set of worksheet questions individually. Attending members of the public were consistently so engaged in small group discussion that they were reluctant to stop their conversations. It was common for some groups to begin their discussions even before the facilitator gave the green light for that portion of the workshop to begin. A combination of factors appeared to be responsible for participants' reluctance to end small group discussions: there was too much ground to cover, they did not have enough time, and/or they enjoyed the conservation and did not want to stop.

Participants reported that this type of interaction was most rewarding. In addition, Forest Service employees came to realize that it was the development of relationships within the small group discussion format that this whole process was all about (personal communication, May 1995).

One difficulty with the small-group format is the strong effect the choice of where to sit has on what kind of exchange takes place and what kind of conversation is created. Attendees often sat down with people they already knew, or with family and friends with whom they were attending the workshop. Usually participants arrived with one or two other people, and there were usually five—although sometimes only two or three—other people at a table. This may have limited the degree to which relationship building actually took place among people who were not already acquainted with each other, because some people sat and talked primarily with people they already knew, and with whom they sometimes already shared a great deal in common.

Finally, a sense of shared purpose, community, and motivation to continue working together as a group was tangible during the last one or two meetings. Several participants made a point to encourage the whole group to continue to work together, noting that they had much more power in working collectively than in individual efforts. In group sharing as well as on written evaluation sheets, several participants stated that in discussions with family and friends, they had encouraged others to participate with them. In the words of one participant, "If you want something done, get involved. Don't complain to me" (May 20, 1995 workshop evaluation).

Conclusion

When viewed against the backdrop of traditional public participation processes, it is clear that the Cispus AMA's 1995 effort was substantially different. The social communication framework of analysis developed here shows several important reasons why. Openness to multiple sources of knowledge, a commitment to mutual responsibility for efforts requiring mutual investment and effort, and recognition of interpersonal relationships as the foundation for collaborative work emerged as crucial features in this study when a social communication approach was used to examine an example of public participation.

The Cispus Adaptive Management Area's first foray into working with the public during the spring and summer of 1995 reflected a substantial departure from the hierarchical, noninteractive approaches to public involvement that have often characterized federal resource management agencies' efforts in the past. The implementation guidelines for the AMAs created an atmosphere of openness to the perspectives of the many parties concerned about and engaged in ecosystem management issues on the AMAs. The Cispus AMA responded to this mandate by engaging citizens in an approach to public participation that was grounded in systems thinking by asking to withhold judgment until they had heard each other out, and by making a sincere effort to ensure that everyone present had the opportunity to speak. Introductions of the AMA to the public at the initial workshop and via the AMA newsletter emphasized the local Forest Service's commitment to an open stance to the multitude of perspectives the public would bring to the planning process.

The thorniest issue that arose when the social communication perspective developed here was brought to bear on an instance of public participation that was associated with control and shared responsibility. Again, federal guidance documents are instructive regarding overall agency approach. In this case, policy governing Forest Service implementation of public participation in AMAs is characterized by a dual charge to develop collaborative modes of interaction with the public while retaining ultimate authority to make decisions within the agency itself. The case of the Cispus AMA revealed that these two mandates are inherently contradictory because maintaining the "locus of control" (McLain, 1995) in one party compromises the level of ownership and investment on the part of the others. In particular, the absence of an opportunity for the entire group to discuss the ideas they generated individually and in small groups prevented the development of an integrated vision that would have been the enactment and outcome of shared responsibility. After this phase of public participation was completed, a couple of citizens noted this gap by commenting that "there was never sufficient time for thought development in response to the exercises, let alone a true interaction and sharing of those thoughts. . . . We were . . . kept so busy with predetermined tasks that there was no opportunity for innovation in the process on the part of the group itself" (personal letter to Cispus AMA, September 1995). Although the issue of control does not alter the fundamentally co-constructed or collaborative nature of communicative interaction—including forums for public participation—when viewed from the perspective of social communication, it is important to note that the collaboration that takes place in the context of a significant power differential is radically different from collaboration in an environment where ultimate decision-making power is decentralized and vested in group interrelationships.

A final example of the Cispus AMA case study helps to underscore this point. In January 1996, the local Forest Service involved the public in a Landscape Analysis and Design (LAD) process that would provide a mechanism to integrate data regarding the condition of the ecology of the AMA with the concerns and values of the public as compiled in the "Cispus AMA Strategies" document that was published in October 1995 as a result of the spring and summer collaborative learning workshop.[10] General feedback surrounding the LAD process was very positive from both Forest Service and the public. Yet when the LAD process suggested a lower timber harvest level than expected, the Forest Service took the decision-making process back in-house. In a May 2, 1996 letter to the participating public, Margaret McHugh stated that:

> The Landscape Design developed through the interagency team and our public is a draft and can be modified. . . . It is now our task to use the information from the analysis to tighten and refine both the lines and definitions of landscape units, while still meeting the intent of the draft Landscape Design. We must also determine criteria to evaluate comparisons between proposals, including how well a plan meets the Cispus AMA Strategies. (letter to Friends of the Cispus AMA)

The team identified in the letter to participate in the definition process included only agency employees. Thus control over deliberations on how to balance environmental and social concerns on the Cispus AMA came back to rest with the agency. Although public comment into the LAD planning process was welcomed, the forum for obtaining public suggestions and concerns clearly reflects a return to a bureaucratic approach to communicating with citizens, in contrast to the many aspects of the collaborative learning process that reflected more of a social communication perspective on planning and public participation.

The focus on collaboration in both the federal guidance documents and the method for public participation chosen by the Cispus AMA indicated strong emphasis by both national and local levels in the Forest Service on working together with nonagency entities. The key component of the public workshops—small group conversation—together with the sustained series of six workshops over the course of several months, created opportunities for citizens to talk with and get to know local Forest Service personnel as well as other citizens that were unprecedented in local experience. Working side by side with Forest Service employees helped several citizens to develop a new sense of respect for them as individuals. Again, from a social communication perspective, two or more interacting individuals are by definition engaged in relationship development, regardless of whether those relationships are seen as constructive regarding a particular issue at hand or not. It is the recognition of this feature of interpersonal communication and the building of a public participation strategy around such awareness that the ability to enhance the democratic nature of participatory deliberations can be found. Finally, it is important to note that these three features are interrelated and interdependent. It is in combination that they form a coherent, consistent perspective; their analytic power derives from their interconnectedness. The integrity of a public participation program from a social communication perspective thus depends on the degree to which all three components are present and energetically facilitated. From this social communication viewpoint, then, it is no small matter when administrative guidelines for public participation reserve primary responsibility for agency decision makers. Although acknowledging that it is also no small matter—indeed, it is nothing short of revolutionary[11]—for federal administrative agencies to consider sharing responsibility with citizens in a decision-building framework (Shannon, 1991), this chapter nonetheless concludes that in an approach to public participation fully informed by a social communication perspective, the goals and the process of public–agency discourse would be mutually determined by all participants through dialogue. In the case of public participation on the Cispus AMA, the lack of such shared control due to macro-scale agency constraints may have reduced the level of ownership and investment on the part of the public in the ultimate outcome of the process and compromised future agency efforts to enlist the participation of the public in forest management activities.

NOTES

1. The Administrative Procedures Act (APA) of 1946 is cited by Rosenbaum (1978) as the dawn of the modern era of legislation explicitly requiring the participation of citizens and citizens' groups in the decision-making activi-

ties of administrative bodies. Others note a groundswell of enthusiasm during the 1960s and 1970s for the participation of the public in planning efforts (Sewell & Phillips, 1979).

2. Many of these strategies are documented in the public participation element of the natural resource management literature (Lawrence & Daniels, 1996; Shindler & Neburka, 1997; Shindler, Steel, & List, 1996). Another aspect of this literature highlights the range of perspectives on what public participation is and its role in the administrative workings of a democratic form of government (Cortner & Schweitzer, 1981; Fairfax, 1975; Goldenberg & Frideres, 1986; Tipple & Wellman, 1989; Wellman & Tipple, 1990; Wengert, 1976).

3. President Clinton's Forest Conference in Portland, Oregon on April 2, 1993 was intended to reopen dialogue regarding forest management in the Pacific Northwest that was fractured over the spotted owl debate (see Lange, 1993; Moore, 1993; and Yaffee, 1994, among others).

4. Shannon (1991) offers "decision-building" as an alternative model to guide administrative planning and the participation of citizens in this process (1991). Decision-building involves both citizens and agency staff in creating knowledge and developing shared commitment toward a mutually defined course of action (p. 27).

5. The FEMAT process was a direct outgrowth of Clinton's Forest Summit. The three-month scientific process culminated in the President's Northwest Forest Plan and laid the groundwork for the Adaptive Management Area policy discussed in the following case study.

6. Public participation in resource management is just one aspect of environmental discourse that is receiving increasing attention from communication scholars. Environmental history, literature, education, and media representations are just some of the other subfields examined recently. Notable examples are Cantrill (1993), Collins-Jarvis (1997), Farrell and Goodnight (1981), Hardy-Short and Short (1995), Killingsworth and Palmer (1995), Lange (1990), Moore (1993), Oravec (1981, 1984), Peterson (1991), and Peterson and Horton (1995). This chapter seeks to make both conceptual and empirical contributions to the growing literature linking the fields of communication and natural-resource management: conceptually by applying an explicitly social constructionist approach to communication, and empirically with an examination of a particular public participation process implemented by the U.S. Forest Service.

7. The collaborative learning approach to agency–public interaction in natural-resource management used by the Cispus AMA was developed by Daniels and Walker (1995). *Collaborative learning* blends a systems perspective with alternative dispute resolution strategies (Daniels & Walker, 1994).

8. See Daniels and Walker (1995) for situation map examples.

9. According to Kruger (1996), competitive pluralism entails processes "that avoid bringing parties with differing viewpoints together to deliberate on

a common good" (p. 26), while deliberative democracy "values democratic participation and civic engagement and builds on a conception of shared values, or common ground that emerges through the deliberative process" (p. 27).

10. *Cispus AMA Strategies,* Randle and Packwood Ranger Districts, Gifford Pinchot National Forest, October 23, 1995.

11. A substantial review of the literature on public policy and the evolution and implications of administrative authority is beyond the scope of this chapter. See DeSario and Langton (1987c), Henning (1974), Langton (1978), McLain (1995), Pennock and Chapman (1975), Reich (1988).

CHAPTER 3

COMPETING AND CONVERGING VALUES OF PUBLIC PARTICIPATION: A CASE STUDY OF PARTICIPANT VIEWS IN DEPARTMENT OF ENERGY NUCLEAR WEAPONS CLEANUP

JENNIFER DUFFIELD HAMILTON

Among the pressing environmental policy issues in the United States today is the environmental remediation of the Department of Energy (DOE) nuclear weapons complex. The nuclear weapons complex includes more than 16 major facilities across the country as well as thousands of smaller support facilities that manufactured nuclear weapons for defense programs beginning in World War II. Weapons development and production operations at these facilities were driven by the nature of the nuclear arms race, which resulted in hasty production schedules and neglect of safe waste-management practices. By the late 1980s, DOE had scaled back production and was faced with enormous hazardous waste management and environmental contamination problems. Many of these facilities have ceased production and are currently undergoing remediation according to the Comprehensive Environmental Response, Compensation, and Liability Act (CERCLA) law, carried out by the DOE, and regulated by U.S. and state Environmental Protection Agencies (EPA) (US DOE, 1995b; 1997).

The emergence of DOE programs that involve the public in environmental management and remediation decisions at these sites offers important lessons about the successes and limitations of government attempts to involve citizens in policymaking. In the late 1980s and early 1990s, diminishing Cold War concerns brought a transition within DOE in which a once self-regulated agency protected by a national security mission learned to deal openly with the public concerning matters that had been top secret since the Manhattan project. Like many government agencies, DOE has developed public programs in an atmosphere described by Laird (1989) as a "decline in deference" (p. 543) or public alienation and distrust of government institutions. In response to increased administrative decision-making authority and reliance on scientific experts for decision inputs, citizens have challenged the accountability of agency officials and become "increasingly reluctant to defer important decisions to institutional elites" (Laird, 1989, p. 543). Fiorino (1996) describes this situation as creating a "participation gap" (p. 194) or a gap between citizen expectations and their actual ability to contribute to decisions.

Government agencies such as the DOE respond to citizens in part by relying on public participation mandates included in environmental laws such as CERCLA or the National Environmental Policy Act (NEPA). At times, these mandates amount to little more than a means for disseminating information and eliciting citizen reactions to proposed plans. For example, CERCLA requires public input after a preferred cleanup method has been selected rather than during the site study and risk assessment stages of the decision process (Applegate, 1998; National Research Council [NRC], 1994). These forums limit participation by constraining participants' ability to influence decision outcomes and often further dichotomize already opposing positions (Killingsworth & Palmer, 1992).

An important source of conflict in these situations is the competing definitions of risk (Slovic, 1987) held by scientists, citizens, and policymakers that broadly can be described as technical and cultural understandings of risk (Plough & Krimsky, 1987). An equally important source of conflict is competing definitions of participation that reflect the gap described by Fiorino (1996) between expectations and ability to participate. This *participation gap* can be seen in the persistent call for mechanisms that allow citizens to not only be informed about issues, but to define and deliberate issues with policymakers. This chapter explores various expectations for public participation as a factor that can enable or constrain public participation processes.

Specifically, this chapter examines ten in-depth interviews with public participants involved in the environmental remediation of one of the DOE's former nuclear weapons sites, the Fernald site located at Fernald, Ohio. This interview data was collected as part of the Fernald Living History Project, a community-based effort to record and preserve the personal narratives of residents and workers regarding ways in which the Fernald site has affected surrounding communities. I explore accounts of public participation offered in the personal narratives and develop themes regarding views or expectations of public participation expressed by these individuals. The themes are compared to four perspectives regarding collective decision making to determine if the conflicting demands identified by the competing-values theory (Quinn & Rohrbaugh, 1981, 1983; Reagan & Rohrbaugh, 1990) are also characteristic of public participation. Ultimately, exploring these discourses will inform the design of inclusive and open processes.

PUBLIC PARTICIPATION LITERATURE AND THE "PARTICIPATION GAP"

Within both scholarly and practitioner literatures on public participation, the need for participatory mechanisms that not only inform interested parties but also allow them to probe, analyze, and debate is prominent. This can be seen in a shift away from one-way forms of communication such as review-and-comment periods or public hearings and toward two-way communication and deliberative processes offered by public meetings or workshops. The more progressive of this literature argues not only for two-way communication and sufficient information but also for sustained, regular interaction among affected parties that allows them to jointly define problems and evaluate alternative solutions. A number of theoretical approaches and participatory models address the participation gap by identifying criteria that would enable public processes to overcome this gap with meaningful participation. These approaches promote democratic participation (Fiorino, 1996), participatory analysis (Laird, 1993), two-way communication (Juanillo & Scherer, 1995), collaborative learning (Daniels & Walker, 1996), and fair and competent processes (Webler, 1995) as the foundation for participation. Further, recent research has turned to the participants to provide definitions of good public processes (Tuler & Webler, 1999).

The participation gap is addressed in democratic models by emphasizing the need for public processes grounded in direct participation in which citizens have the opportunity to "co-determine" decisions with agency officials (Fiorino, 1989a; 1996). According to this perspective, a decision-making process that considers citizens as partners and values experiential forms of knowledge is grounded in four democratic criteria: direct participation,

participation on an equal basis, face-to-face interaction over time, and the ability to share in decision making (Fiorino, 1989a). The democratic concern with open mechanisms includes a focus on citizen access to relevant information and agency officials as well as opportunities to develop understanding and group learning around the issues. Such participatory analysis is based in activities of learning and analyzing issues to make informed judgments and is the necessary foundation for involvement in complex technical issues such as environmental policymaking (Laird, 1993).

Several communication scholars have envisioned constraints on public participation by contrasting one-way versus two-way models of communication between the public and decision makers. One-way communication is seen as an informational tool used to persuade the public to accept a scientific assessment. In contrast, two-way communication facilitates an exchange of information and opinions by including multiple perspectives, a free flow of information, and access to communication channels and resources (Juanillo & Scherer, 1995). This basic contrast between a one-way transfer of information from experts to citizens and a two-way exchange of information and values that informs decisions is seen in most communication models that support a democratic ideal of participation. However, an important distinction can be made between a two-way communication model in which technical experts share scientific knowledge while the public shares community values and a model in which there is a two-way transfer of both facts and values in both directions (Waddell, 1996).

Further, public participation is not simply the transmission of information—whether in one or two directions—but a process of social construction and negotiation of information (Graham, 1997; Katz & Miller, 1996; Rowan, 1994). In the collaborative learning approach developed by Daniels and Walker (1996), participants are not simply exchanging information, but are also engaged in a "process of defining the problem and generating alternatives [that] makes for meaningful social learning as constituencies sort out their own and others' values, orientations, and priorities" (p. 73).

The democratic impulse to resolve public affairs through a debate of opinions and the emergence of consensus is the basis for a final approach focused on criteria that enable participation. Webler's (1995) normative theory of public participation is grounded in the Habermasian vision of an undistorted communication capable of reinvigorating the public sphere. Webler envisions fairness and competence as meta-principles for public participation. Fairness is conceptualized as providing citizens the ability to attend public forums, initiate discourse, participate in the discussion by asserting and challenging claims, and participate in decisions for both agenda setting and the issue itself. Competence is conceptualized around criteria that allow the emergence of understandings based on accessible information.

Several of the models mentioned here were derived from democratic, communication, and critical theory. According to researchers Chess and Purcell (1999), such "universal goals and criteria derived from theory are less important than the specific goals of those involved in participatory efforts" (p. 2686). In this spirit, the recent work of Tuler and Webler (1999) has focused explicitly on definitions of a good public participation process expressed by participants of a forest policy discussion as the basis for seven principles of public participation.

This chapter contributes to this work by exploring participant views regarding public participation and considering the possibility that underlying values about public participation contribute to the conflict or consensus that emerges through public processes. Here, I envision public participation as a collective decision process and turn to a conceptual framework that looks at ways in which group decision processes may be evaluated according to competing values.

PUBLIC PARTICIPATION AND COMPETING VALUES

The "competing values approach" developed within the organizational literature is based on the idea that there are competing values concerning the purpose of group decision making and that they place conflicting demands on the process (Quinn & Rohrbaugh, 1981, 1983; Reagan & Rohrbaugh, 1990). This approach has been applied to public participation as a decision-making process (Hamilton, 1997; Mumpower, 1995; Vari, 1995). This chapter argues that participants in environmental issues often have expectations about the purpose, structure, and outcomes for public participation that can be envisioned according to the four perspectives of effective decision processes outlined in the competing values approach.

The competing values approach is based on the notion that "the criteria commonly used to assess collective performance effectiveness reflect alternative priorities for any group or organization" (Reagan & Rohrbaugh, 1990, p. 24). The alternative priorities or values are conceptualized as tensions between the need for flexibility and the need for control, a focus on internal group issues and a focus on external contexts, and a focus on process as a means or an end. Various combinations of these three values are envisioned in this approach as presenting competing expectations for group decision processes that result in four perspectives on group process effectiveness: the rational perspective, political perspective, consensual perspective, and empirical perspective. The four perspectives reflect competing values because "they emphasize what often appear to be conflicting demands on the decision-making process" (Reagan & Rohrbaugh, 1990, p. 25).

A rational perspective is one that values a goal-centered process. Persons who view collective decision making from this perspective are focused on external concerns and value a controlled, task-oriented process for reaching decisions. The rational perspective views collaboration as a means to achieve a goal and values "methods that efficiently assist decision makers with improving the consistency and coherency of their logic and reasoning" (Reagan & Rohrbaugh, 1990, p. 26). The rational perspective views group interaction as a *goal-centered process* and is concerned with *efficiency of decisions*.

The political perspective is also externally focused, but values a flexible process for making decisions. Collective decisions are viewed as an end goal in and of themselves because participation brings legitimacy to the decision. Idea generation and creative problem solving facilitate the best decisions possible, according to this perspective. The political perspective views that "the search for legitimacy of the decision and how acceptable it is to outside interest groups and external stakeholders would be notable through a fully adaptable process" (Reagan & Rohrbaugh, 1990, p. 26). The political perspective views group interaction as an *adaptable process* and is concerned with the *legitimacy of decisions*.

The focus in the consensual perspective is on internal group dynamics and interpersonal interactions. A flexible process of team building and idea sharing is valued. This perspective expects full participation in discussions that create an environment in which thoughts and ideas are freely expressed. The consensual perspective values "extended discussion and debate about conflicting concerns [that] should lead to collective agreement on a mutually satisfactory solution" (Reagan & Rohrbaugh, 1990, p. 26). The consensual perspective considers group interaction as a *participatory process* and is concerned with the *supportability of decisions;* therefore, the process is seen as an end and not a means.

The empirical perspective is also focused on internal concerns but favors a controlled process. From this perspective, decisions should be made based on collecting all the information and data available on the problem. Participation in group decisions is a means of eliciting all relevant information needed to make a decision. This information is used to develop "large and reliable databases to provide decision support . . . [and] thorough documentation and full accountability" (Reagan & Rohrbaugh, 1990, p. 26). The empirical perspective considers group interaction as a *process of data collection* and is concerned with the *accountability of decisions*. The differences in these perspectives are summarized in Table 3.1.

These four perspectives express different values and expectations about collective decision making and provide a useful way to examine participants' expectations for public participation as a decision process. In the sections that follow, I present a case of sustained interaction among the DOE, the U.S. and Ohio EPAs, and neighbors of the former DOE nuclear weapons facility located in Fernald, Ohio. The views of public participation held by participants involved in remediation decisions for the Fernald site are analyzed to explore the ability of this approach to extend our current understandings of the factors that enable and constrain public forums. The analysis will focus on two questions: (1) Does this conceptual framework portray the ways in which participants assess public participation? and (2) Can the conflict or consensus that emerge from public interaction be explained in part by differing or overlapping expectations for public participation? Before turning to the voices of participants to explore these questions, the case study to be used in this analysis follows.

Table 3.1
Summary of Differences in Perspectives

CVA Perspectives	Internal Focus	External Focus	Flexible Structure	Controlled Structure	Collaboration as End	Collaboration as Means
Rational Perspective "efficient decisions"		X		X		X
Political Perspective "legitimate decisions"		X	X		X	
Consensual Perspective "supportable decisions"	X		X		X	
Empirical Perspective "accountable decisions"	X			X		X

Note. From "Group Decision Process Effectiveness: A Competing Values Approach," by P. Reagan and J. Rohrbaugh, 1990, *Group & Organization Studies, 15*(1), pp. 20–43.

THE FERNALD NUCLEAR WEAPONS SITE

In the mid-1980s a transition took place in the rural farming communities surrounding the Fernald nuclear weapons site when it was publicly announced that they had hosted a government facility that processed uranium for nuclear weapons. The Fernald site was part of the DOE nuclear weapons complex and it manufactured uranium products from 1953 until 1989 when the facility ceased production after environmental pollutants were found in nearby land, air, and water. Since 1989, the site has been under environmental remediation directed by the U.S. and Ohio EPAs according to CERCLA law. Production and waste management practices at the site were driven by the nuclear arms race, national security concerns, and a culture of secrecy in which workers operated under a need-to-know mentality and neighbors were largely unaware of the nature of operations at the plant. Protection of the environment and human health were secondary to the Cold War effort (US DOE, 1995b; US DOE, 1997).

In 1984, after three decades of secrecy on the part of the government, DOE hosted a public meeting at a local elementary school to announce that 300 pounds of enriched uranium had been released from the plant into the atmosphere. At this meeting, it was also announced that in 1981 uranium had been discovered in three off-site water wells of neighbors (US DOE-FEMP, 1995). These announcements evoked concern in neighbors whose drinking-water wells had been contaminated with uranium and in some plant workers who were potentially exposed to hazardous materials. The public participation that has developed in this community has resulted from a combination of grassroots efforts, DOE programs, and a site-specific advisory board; it is considered one of the most notable instances of civic involvement in the nuclear weapons complex (Keystone Center, 1996).

A grassroots organization developed in 1984 under the name Fernald Residents for Environmental Safety and Health (FRESH). They have been a constant presence in Fernald decisions over the last seventeen years and have spent countless hours studying CERCLA documents and attending and commenting at DOE public meetings. FRESH has struggled with DOE to clean up the site, compensate its neighbors and workers, and protect the health of the community and local environment (US DOE-FEMP, 1995). The relationship between FRESH and DOE has evolved from litigation and antagonism in the early days to one of cooperation and involvement. FRESH's early participation revolved around advocacy efforts to raise awareness about health issues and demand a voice in DOE cleanup decisions. In the early stages of participation in the mid-1980s, citizen activism was critical to drawing agency attention and fostering openness to citizen perspectives. Over time their participation has shaped the development of public programs by DOE and its contractors.

Though the relationship between DOE, regulators, and Fernald-area communities began with contentious discussions and litigation, it has developed into one of ongoing and cooperative interactions. CERCLA and the Resource Conservation and Recovery Act (RCRA), as the primary laws guiding Fernald remediation, stipulate public involvement activities that must take place as the site accomplishes various phases of CERCLA remediation such as public notifications, comment periods, and meetings (US DOE, 1991, p. 18). In addition to these legal requirements, the public program at Fernald includes supplemental activities that follow DOE guidance (US DOE, 1991) and several innovative approaches designed as part of the site's 1993 public involvement program (US DOE-FN, 1993). For example, through the "envoy program," site employees serve as liaisons to civic groups throughout the community. Additionally, DOE and its site contractor, Fluor Daniel Fernald (FDF), have emphasized the need for management to engage the public and develop channels for face-to-face contact that establish personal relationships with members of the community. This has evolved to include informal envoys and multiple points of contact that create a network of core community and opinion leaders in the area.

The second key group is a citizens' advisory board for the site that was instrumental in providing advice on crucial decisions about site remediation. The site-specific advisory board was convened in 1993 as the Fernald Citizens Task Force (FCTF) and was later renamed the Fernald Citizens Advisory Board (FCAB). A national discussion group had recommended a model to improve DOE public participation efforts based on the idea of a citizens group interacting over time, studying and debating the issues, and having access to high-level management and necessary technical information (Keystone Center, 1993). At this point in time, Fernald was in the CERCLA phase of selecting a remedy and realized the benefit of such input. DOE officials at Fernald, therefore, convened the FCAB and asked them to provide consensus recommendations on four key issues impacting site cleanup: future use of the site, cleanup levels, cleanup priorities, and waste disposition options. Key to the success of the FCAB was its incorporation of wider community input in its deliberations through workshops and informational efforts (FCTF, 1995). The FCAB submitted its recommendations on these four crucial issues in 1995 and continues to be a central means for generating ideas about the cleanup effort and long-term stewardship issues.

Therefore, the successes at Fernald are due in part to the dedication of residents who have remained resolved to find a satisfactory solution through public means. They are also due to DOE and contractor personnel who reached beyond directives and legal requirements to design a participation program for Fernald. Interviews with ten people who have been actively involved in participation efforts at Fernald follow and are analyzed to determine their expectations for public participation and factors they found key to forging a collaboration.

DISCOVERING PARTICIPANTS' PERSPECTIVES

This chapter is an exploratory analysis of selected interview data collected as part of the Fernald Living History Project, a community-based effort that represents the collaboration of community leaders, a local historical society, the Center for Environmental Communication Studies at the University of Cincinnati, DOE and its contractor Fluor Daniel Fernald (FDF), and Ohio EPA. The project began in October 1997. Over the course of the last several years, project partners conducted interviews to create a historical archive of personal narratives of community and worker experiences with the site. Project activities are intended to promote a broader public understanding of the Fernald site's social legacy and significance for future public policy.

The archive includes a broad range of community perspectives including area residents, site workers, government officials, and members of organized groups. Interviews have been conducted in three phases. An initial round of six pilot interviews was conducted in the summer of 1998 to develop interview protocol and produce a promotional video to encourage neighbors and workers to share their experiences through the project. A second round of 107 interviews was conducted by DOE and FDF during 1999. Finally, in 2000–2001, a round of 25 interviews was conducted by the Center for Environmental Communication Studies under a grant from Ohio EPA's Environmental Education Fund. The author has been involved in this project from its inception and has served as interviewer for 12 of these interviews. These stages of interview collection form the historical archive that has been developed to date with this community.

The interview approach followed an informant-directed style in which the researcher relied on open-ended questions and prompting and floating questions, but the conversation is primarily guided by the informant (McCracken, 1988). This approach is intended to provide as little structure as possible to allow the interviewee's definition of the situation and view of what is relevant to drive the conversation and shape the data to be yielded by the interview. Informant-directed interviews form "a mixture of conversation and embedded questions . . . [in which] the questions typically emerge from the conversation" (Fetterman, 1998, p. 39) and represent a combination of descriptive questions that ask about key events and evaluative questions that probe into meanings they ascribe to those events (Whyte, 1984). This approach allows the researcher to uncover cultural meanings within the informant's discourse that might otherwise be missed if the conversation was more managed or structured (Fetterman, 1998; Peterson et al., 1994).

A preinterview was conducted over the telephone with each interviewee to familiarize him or her with the interview style, establish rapport, and develop a sketch of his or her experiences with the site. The information obtained in the preinterview served as the basis for prompts and follow-up questions used during the interview. Interviews lasted anywhere from 1 to 3 hours depending on how much the informant wanted to say during the interview. The interviews were videotape recorded in the interviewee's home, office, or another location of their choosing to allow for a natural setting in which they felt most comfortable. Transcriptions of each interview are included in the historical archive.

The unstructured nature of this interview approach is appropriate for eliciting personal narratives because it allows interviewees to define and articulate the important elements of their experience and provides insight into the historical perspectives, memories, and experiences of Fernald neighbors, workers, and government officials. McCracken (1988) describes this interview method as one that can "take us into the mental world of the individual, to glimpse the categories and logic by which he or she sees the world" (p. 9). An informant-directed approach has been used in environmental communication studies to investigate views of risk and safety within the meaning systems of a farming culture (Peterson et al., 1994) and myths operating within Texas ranching culture that influence ranchers' views of property rights and wildlife habitat preservation (Peterson & Horton, 1995).

From the historical archive produced through the Fernald Living History Project, 10 interviews were selected based on the active and sustained involvement of these individuals. This included three site-specific advisory board members, three grassroots organization members, two Ohio Environmental Protection Agency officials, and two Department of Energy officials. Interview transcripts and videotapes of the 10 interviews provided the data for analysis.

The method of analysis for this chapter was based on a grounded approach in which relevant themes representing the meaning system of the interviewees emerge from the data through a method of coding and comparison (Glaser & Strauss, 1967). I reviewed the transcripts and videotapes of each interview and developed codes representing descriptions of public participation. These codes were then related to one another to develop themes for each individual interview. The themes were compared across interviews to develop patterns that represent views of participation for the group of interviewees. This approach includes a concern with grounding themes that are developed in the discourse of interviewees and is one of the most effective ways to understand how people assess their world (McCracken, 1988; Peterson et al., 1994; Peterson & Horton, 1995).

PARTICIPANT VIEWS OF PUBLIC PARTICIPATION

The participants described a transition in the relationship among site officials, regulators, and neighbors that took place in the early 1990s.[1] This transition from hostility to cooperation was described by an Ohio EPA official as: "You had a lot of old school DOE, which was secrecy and silence and announce and defend and a few new guys around saying maybe that ain't the way we should be doing it . . . so it's been a significant transition in the DOE folks." Similarly, a FRESH member explained: "in the beginning and early years we called it the decide-announce-defend mode of DOE . . . there were no discussions on anything. [Now] we do really have a seat at the table. You know the stakeholders and FRESH have a seat." In describing this time period, participants attributed the transition to a series of legal actions brought by Ohio EPA against DOE for natural-resource damages and to a new regime of DOE and contractor personnel. These factors, coupled with the end of the site's production mission, led DOE to realize that they had an obligation to clean up the site and to respond to the grassroots groups and regulatory community that were gaining entree into its decision process.

Most accounts of public participation in the 10 personal narratives analyzed here revolved around this transition. These descriptions provide insight into how participants define key elements of public participation. In describing the cooperation that has developed, participants emphasized the following four themes: interpersonal relationships, open and ongoing communication, a learning process, and credible outcomes. These themes each reflect elements of the four perspectives on decision making described in the competing values theory.

INTERPERSONAL RELATIONSHIPS

The most prevalent theme in these personal narratives was that neighbors, advisory board members, and Ohio EPA and DOE officials have developed interpersonal relationships that have enabled this transition. Key concepts that were part of this theme were trust and respect, which participants saw as the foundation for building interpersonal relationships. These relationships were described as necessary to engage in productive group interactions and develop agreement. A DOE official recalled that "until we got a level of trust between us and them, we really couldn't get very far. And it took, I think, a couple of years." A FRESH member elaborated on this point, saying:

> We have all gained respect for each other, it has been, we have learned from each other, we have taught DOE a lot, how to respect the general public's opinion. We can make decent decisions. . . . It took a while for all of us to get together, but public participation has helped us get to know one another, get to respect one another, especially for the DOE and the contractors to realize that the general public can make decisions and we are not always on the opposite side. And, by working together and sitting down at the table, we have gotten respect for each other.

This description emphasizes that the trust building involved not only citizens learning to trust DOE, but DOE had to trust the neighbors as well:

> [A]nd there again was the trust factor. They were afraid to let you see a draft document because you would say what's in it is gospel. And once they realized we understand what draft means, you know . . . they found out we understood and weren't going to hold them to that, and that we could work on stuff. It helped a lot.

An FCAB member commented that trust and one-on-one relationships were developed through gathering information they needed to make recommendations. These relationships enabled a cooperative effort among the board, DOE, and EPA in jointly generating and evaluating options for the recommendations. He explained that DOE and EPA officials served as ex officio members of the FCAB. This permitted board members to question and interact directly with officials: "[We] began to develop relationships with these people who were making the decisions or developing the information that was used in decisions." By asking questions and probing for information, they not only developed relationships, but also began shaping what data was generated for decision making. Another FCAB member described that this led to a process in which all viewpoints were a part of formulating recommendations, which made them supportable decisions. He explained:

> [W]e all kind of worked together on coming up with these recommendations. So that because DOE was part of the process . . . we would ask them questions and get our responses well in advance of formulating our recommendations, and the regulators the same way. And so since all of us worked cooperatively to come up with the recommendations they stood a very, very good chance of being accepted. That chemistry was very, very important.

This statement reflects a consensual perspective and the view that full participation and extended discussions led to supportable decisions. A FRESH member commented that the existence of a relationship also led to the ability to get information more easily, saying:

> [A]ll sides started to have a little more trust and to work together a little bit more. Things just went so much smoother once that happened. . . . You knew you could pick the phone up and call and if they weren't in they would return your call and answer questions. Where before you couldn't talk to anybody.

Participants described these face-to-face encounters as improving participation by allowing them to know the people who generate decision information. The theme of developing trust and interpersonal relationships was dominant throughout their discourse as participants explained that cooperation is needed to foster full participation and group interaction in which participants feel free to discuss ideas and develop solutions. Participants' regard for this dynamic reflects a consensual perspective on decision making because interpersonal relationships were depicted as the key to enabling productive deliberations that incorporate all viewpoints. This theme is related to the next theme of fostering open and ongoing communication through those interpersonal relationships.

Open and Ongoing Communication

The participants pointed to the development of openness and candor with the public as central to the transition in the relationship between the DOE and surrounding communities at Fernald. FRESH members recalled that when they filed a citizen lawsuit against DOE in the mid-1980s, "we felt like we had no other choice than to do that, because nobody would talk to us, nobody would answer our questions." An Ohio EPA official explained that a change came about partly because FRESH argued over and over: "why hide stuff, why not just tell us? If it's bad news, tell us. If things aren't working right, tell us. And I think eventually that got through and DOE started being very open." This has resulted in "communication that is wide open. . . . If the site has a problem FRESH gets a fax. A fax of a daily incident report or something like that." It can also be attributed to DOE's attempt to make information more accessible by creating a network of contacts and channels of communication to the site. A DOE official described the envoy program previously mentioned as providing: "anybody who had a concern for this site has at least one person that they know they could call up on the phone and say 'what the heck is going on' and that person could give them a reliable and trustworthy answer."

The participants also described interaction among the public, site officials, and regulators as taking place on an ongoing, almost daily basis. Rather than restrict input to formal, legally mandated points in the process, an Ohio EPA official commented, "I think that's a tribute to DOE and the contractor's efforts to bring the public into the decision making and involve them in the day-to-day and the week-to-week and important decisions at Fernald." An FCAB member explained: "Now if anything happens the stakeholders are contacted personally at the very beginning, probably even before the media is alerted. So the whole idea of openness and education has changed dramatically over the last several years."

The impact that daily interaction has had on the decision-making process was explained by a FRESH member. She described a shift from a time when ". . . the very first documents that were done, the response summaries are probably almost as thick [as the document] because of all the public comments that they would have to respond to" and toward a situation in which "they brought you in from the beginning and you were involved in making the decisions. There were a lot less comments and stuff because you kind of worked everything out before you got to that point." In other words, the day-to-day interactions among Fernald participants have resulted in an ongoing deliberation of the issues in which concerns are worked out in advance of a proposed solution. This allows citizens to influence the formulation of solutions rather than simply react to agency solutions with formal comments. It might also enable them to be more satisfied with the decision options that are considered by the agency.

Participants described productive public participation at Fernald that has resulted from an approach in which the public is openly told about issues as they emerge and is engaged in a continuous discussion in which their ideas are incorporated into decisions as they take shape. This view reflects a blend of consensual and empirical perspectives on decision making. The emphasis on a continual dialogue and evaluation of ideas is indicative of a consensual perspective's focus on extended discussion and free expression that lead to a solution that satisfies all parties. The concern with open communication as providing early and complete access to information reflects an empirical concern with generating all the data needed to make decisions that are accountable.

LEARNING PROCESS

The third theme, prevalent primarily in the discourse of grassroots and advisory board members, was that an appropriate learning process is essential to making informed decisions. These two participant groups in particular said that such a learning process requires a basic level of information and appropriate analysis tools. For example, an FCAB member suggested: "I think the most important thing we did was, was education really. . . . For one thing, the people on the Task Force had a lot, had a wide variety of knowledge about the site . . . and so, we had to get everyone up to some basic level of knowledge." For the advisory board, the common base of information included technical information and agreement on consensus values that they relied on to evaluate various options and develop recommendations. FRESH members, too, described basic knowledge as a necessary foundation for providing input on decisions, recalling: "FRESH educated themselves on this. Attended many meetings, asked DOE to have meetings in areas that we were not sure of and we need more information."

Another key aspect of the participants' learning process was the use of educational tools that allowed them to collect information and to visualize and analyze the issues. Two tools in particular stood out as important in participant descriptions: the cancer map and the FutureSite game. In the early days of public participation at Fernald, several FRESH members developed a map known as the "cancer map" with the Fernald site in the center, circles surrounding it at distances up to a five-mile radius, and pins marking cancer cases within those areas. FRESH members have used this map as a mechanism for collecting information about cancer cases within their community. "So we thought about the map and trying to show a pattern of . . . possibly of . . . the people who were affected," explained a FRESH member. She continued describing the cancer map as an:

> educational tool. . . . We were trying to get health studies and have CDC or even our public health department to get involved, and so that's why we developed the map thinking that if there was enough information and . . . people would start to see this, that they would start asking for help from these agencies.

The other analysis tool described by participants was FutureSite, a board game like monopoly that the FCAB developed to analyze various cleanup levels and future use scenarios for the site. Using this game, FCAB members moved chips representing waste on site to see the volumes of waste, transportation alternatives, disposal alternatives, and costs required to achieve various levels of "clean" needed to attain specific future uses. An FCAB member commented:

> We had a couple of open meetings for residents and others to come in and play the game. We had one for site managers, one for students. . . . That was very helpful; people began to understand the complexity, the technical complexity of the cleanup of the site, the volumes of stuff that was going to have to be moved if you were going to clean it up to certain levels.

Another board member elaborated on the impact of the game, saying:

> And so we had a three-dimensional visualization of how much material are we talking about and it blew us away. The amount of material that would have to be removed to move everything would be tremendous . . . and that this site would be actively moving material for decades.

A majority of the participants described the FutureSite game as leading the FCAB to a balanced approach for the disposal of Fernald waste. This balanced approach was part of an FCAB recommendation that the site build an on-site disposal facility to dispose of Fernald's low-level waste while transporting the high-level waste out West for burial. FRESH and FCAB members depicted this as a tough decision that ran counter to the typical NIMBY (not in my backyard) response of communities. They conveyed pride about their decision to accept responsibility for some of the waste rather than shifting the problem to another community. A FRESH member illustrated this sentiment: "Consciously we decided that this would be our burden that we would bear. This would be our little piece of the nuclear weapons age and the Cold War. And you know it wasn't an easy decision, but to me it was the right decision."

Participants explained that a learning process that provides complete information and analysis tools is needed to make informed and accountable decisions such as their recommendation to adopt a balanced approach for waste disposal. For example, an FCAB member suggested that a major accomplishment of the board was ". . . dealing with complex and technical issues and trying to understand them to the point where you could make a reasonable recommendation." A FRESH member declared: "We read documents until we were blue in the face. And we checked facts and we double-checked facts and um, more than I ever wanted to know. But in order for us to make good conscious decisions we felt like we had to do that." The sense among FCAB and FRESH members was that they had to become informed to provide accountable feedback. They expected the outcome of their deliberation and analysis to be conscientious and informed decisions.

The learning process that participants described as essential to successful participation embodies a mixture of consensual and empirical perspectives. An empirical view is reflected in the importance they placed on informed and accountable decisions. However, they consider such decisions as requiring not only complete information, but also a deliberative process of extended debate resulting in agreement. The consensual perspective is thus reflected in the importance they place on deliberative processes.

Credible Outcomes

A final theme that emerged from participants' discourse was a concern with a credible outcome of the public process. FRESH and FCAB participants described successful outcomes in terms that parallel a consensual view. They conveyed the sense that the outcome of collaboration was successful if DOE incorporated or considered their feedback in making decisions. This point was underscored by a FRESH member who described that a major milestone for public involvement was:

> The records of decision all reflect what FRESH pushed and harped and we battled out in workshops or round tables. . . . They basically took the input they got from the public meetings and the CAB and FRESH and they did actually incorporate it into the final record of decision on like the on-site disposal cell, the cleanup levels for the soils and aquifer.

In describing the FCAB recommendation for DOE to accelerate cleanup at the site, a board member explained the importance of impacting the final decision saying: "So here was a citizen's group that made a recommendation that in effect altered national policy on the cleanup process."

These participants expressed a view that not all public input would or should be incorporated into policies, but it was important to them to know that their advice was considered and to receive feedback on that advice. A FRESH member noted the importance of DOE listening in her comment: "I don't know if they have always liked what we have come up with, but they have always responded. And most of it they have tried to incorporate in." Similarly, an FCAB member noted: "I don't expect them to accept every recommendation we make and take every decision based upon those recommendations. I expect them to consider them, but if they reach a different conclusion, I'm satisfied for the most part."

In contrast, DOE and Ohio EPA officials expressed the sense that outcomes were credible if the process led to decisions that were acceptable to the broader public and policymakers. The terms they used to describe credible outcomes parallel a political perspective. For example, an Ohio EPA official described the benefit of the FCAB saying: "having this group of interested stakeholders provide feedback to DOE on activities and decisions, I think it provides more credibility to the decisions, both to the DOE side and to the general public's side. . . . DOE is able to use that backing to make decisions and move forward." An example of this backing and credibility assisting DOE decision makers was the FCAB's recommendation to build a disposal cell on site. Several participants noted that this was "a tough sell to this community" that probably would not have been accepted had it not come from a citizens' group.

A DOE official commented that the FCAB has "evolved into making those decisions but also being involved in things like validating or helping set priorities." Another Ohio EPA official suggested that, "Stakeholder involvement is necessary to be successful and to address the needs of everybody. Without it you develop something that probably isn't workable or is certainly going to be fought all the way." These statements reflect a view among the DOE and Ohio EPA participants that public participation plays a crucial role in the decision process by ensuring the legitimacy of decisions to outside groups—a legitimacy that promotes the ability to move cleanup efforts along because the decisions are accepted.

Therefore, this final theme reflected a tension between a consensual perspective held by the FCAB and FRESH and a political perspective held by DOE and Ohio EPA officials regarding successful outcomes. Citizens articulated credible outcomes as decisions that reflect the incorporation or consideration of their input. This articulation reflects a consensual perspective and emphasis on decisions that can be supported by all interested parties because they were involved in formulating them. Agency officials articulated credible outcomes as decisions that are acceptable to the broader community and policymakers. This articulation reflects the political concern with collective processes that imbue decisions with legitimacy.

DISCUSSION AND CONCLUSIONS

Many of the descriptions of successful public participation offered by Fernald participants confirm aspects of the democratic, communication, and critical theory models discussed earlier. Participant concerns with open communication, the ability to influence decision outcomes, and the necessity of learning tools for raising and deliberating technical issues are examples of Fernald participant views that are consistent with criteria offered in the literature. Fernald participants confirmed the importance of interpersonal relationships identified in recent research (Tuler & Webler, 1999; Graham, 1997). In addition, Fernald participants articulated elements of successful participation in terms consistent with the perspectives defined in the competing values theory.

The political, consensual, and empirical perspectives were evident in the descriptions of public participation offered by Fernald participants. In particular, the theme of interpersonal relationships reflects a consensual perspective toward decision making and was strong in all the interviews analyzed here. The theme of open and ongoing communication reflected consensual and empirical perspectives and was similarly evident in all interviews. The need for a learning process reflected consensual and empirical perspectives and was present primarily in the discourse of FCAB and FRESH members. Finally, the theme of credible outcomes was important across participant groups; however, it was articulated differently by citizens and agency officials. FCAB and FRESH members articulated this theme in ways parallel to a consensual perspective whereas agency participants described this theme in terms consistent with a political perspective. It is important to note that the rational perspective was not a part of their vocabulary for discussing participation. This may be a result of the tendency to talk about public processes in ways that downplay efficiency because a task orientation is seen by some as detracting from process aspects of interaction, a point that deserves fuller exploration in future research.

The first question addressed in this chapter was whether the competing values framework portrays the ways in which participants assess public participation as a form of group decision making. The finding here is that it clearly does, and further that there appears to be a blending of perspectives *within and across participant groups*. The ways in which this blending of perspectives impacts expectations for public participation is related to the second question guiding this analysis. Can the conflict or consensus that emerges from public interaction be explained in part by convergent or divergent expectations for public participation among participant groups?

Several preliminary conclusions may be drawn from considering this question. At Fernald, participant views of a successful public process seem to have converged across participant groups as the agencies and neighbors at Fernald have come to think of their collaboration according to a consensual perspective. This was evident in consistent descriptions of the importance of trust, interpersonal relationships, and ongoing interactions in which the public discusses issues as they emerge and citizen feedback is incorporated into evolving decisions. This shared vision of a good process has formed a common vocabulary for Fernald participants and common expectations for elements of a successful process. This shared vision of the values and priorities important to participation may influence the development of agreement within these group deliberations.

An important distinction was evident, however, between views held by FRESH and FCAB members and those of DOE and Ohio EPA officials. The citizen perspective blended empirical concerns of studying the issues and making accountable decisions with a consensual view. In contrast, agency officials blended political concerns of validating or legitimating decisions through a public process with a consensual view. The competing values theory considers these empirical and political perspectives as opposites. The empirical emphasis on internal group dynamics and a controlled structure competes with the political focus on external stakeholders and a flexible process. Further, an empirical perspective sees group process as a means for gathering relevant information while the political sees it as an end because it validates decisions. According to the theory, these alternative priorities may produce tensions that stem from divergent views of process. In this case study, however, Fernald participants did not point to this as a hurdle to their cooperation. Future research should explore whether the competing values of empirical and political views do influence the inability of other groups to reach agreement.

Relating the competing values approach to the themes that emerged from Fernald participants' discourse revealed that this group holds a set of common expectations for participation. The consensual perspective was predominant despite the presence of an empirical view among citizen participants that competes with a political view held by agency participants. This overriding concern with a consensual process may aid their ability to reach agreements; however, the fact that they agree on elements of a successful process does not predict their ability to agree on the substance of issues.

This case study of one instance of public participation has explored the elements of successful process articulated by Fernald participants and has related them to competing expectations for group decision making. Chess and Purcell (1999) argue that neither outcome nor process success can be determined by the form of participation; therefore, researchers must look to other contextual factors to explain why a given instance of public participation worked. Here I have offered a preliminary analysis of why it worked at Fernald by tracing themes in the discourse of Fernald participants. The four themes of interpersonal relationships, open and ongoing communication, a learning process, and credible outcomes articulate elements of successful participation at Fernald.

This chapter offers a preliminary look at how these elements, many of which are consistently identified in participation literature, relate to competing expectations for public processes. Further research should explore the possibility that the distinction between empirical and political views is a source of disagreement during participant interactions because of the conflicting demands they place on the process. The emphasis on interpersonal relationships as a crucial element of successful participation suggests a future direction for literature that examines criteria for successful public participation. Fernald participants pointed to these relationships as equal in importance to democratic or deliberative design elements of various public participation mechanisms. A better understanding of the interplay of interpersonal relationships and democratic aspects of a consensual process would extend our current grasp of the factors that enable and constrain public forums.

NOTE

1. Quotations are taken from transcripts of interviews completed as part of the Fernald Living History Project from 1998 to 2001. Copies available on request.

CHAPTER 4

PUBLIC EXPERTISE: A FOUNDATION FOR CITIZEN PARTICIPATION IN ENERGY AND ENVIRONMENTAL DECISIONS

WILLIAM J. KINSELLA

Energy and environmental issues are among the most consequential in our contemporary "risk society" (Beck, 1992, 1994, 1995, 1999). They are also among the most contested. As sites of what Walter Fisher (1987, 1994) calls "public moral argument" and what Pearce and Littlejohn (1997) call "moral conflict," these issues are marked by deep disagreements among stakeholders with divergent values and rationalities. It is often assumed, by both participants and observers, that these stakeholders' worldviews are fundamentally incompatible, and this assumption only deepens perceived differences. Pearce and Littlejohn describe such conflicts as "interminable," "intractable," "rhetorically attenuated," and "morally attenuated." As the last two of these descriptions suggest, whereas a wide range of arguments and values are relevant and necessary to addressing these conflicts, few of these actually enter into public discourse.

Fisher notes that public moral argument often collapses into a narrow conversation dominated by technical expertise. The diversity of strongly held perspectives makes common ground difficult to achieve, leading participants to converge on technical rationality as the only widely accepted form of argument. As Toumey (1996) puts it, the "plenary authority of science" prevails, displacing other values and interests that have no common mode of articulation. In some cases, expert communities become increasingly influential while the public becomes increasingly alienated, leading to monopolistic decision making with little legitimation. In other cases, public debates reduce to competitions between expertise and counterexpertise as each side deploys sympathetic specialists in an ultimately irresolvable cycle of technical debate (Balogh, 1991; Beck, 1999; Collingsridge & Reeve, 1986; F. Fischer, 2000). In all of these cases, what gets left out is the most essential component of public discourse: the effective expression and exploration of individual and community values. Value-free science is not capable of expressing those values, or of connecting them to the technical questions that it addresses (Proctor, 1991). The consequences are multiple and unfortunate: decisions fail to serve the interests of those most closely affected (Bullard, 1990, 1994), technocratic institutions lose their legitimacy (Habermas, 1975), and democracy is short-circuited (Barber, 1984; Sclove, 1995; Williams & Matheny, 1995).

F. Fischer (2000) has provided a valuable analysis of this problem in the context of environmental policy debates. Drawing on post-positivist and constructivist models of science, Fischer argues that although science is surely essential in the analysis and consideration of environmental issues, it needs to be integrated with the "local knowledge" and "multiple discourses" of affected people and communities. Such integration is required not only to place scientific questions and conclusions in their appropriate social contexts, but also to make scientific analysis more valid and complete:

> Not only are the intentions and motives of the locals essential to a proper understanding of a situation, but they also typically possess empirical information about the situation unavailable to those outside the context. While such local knowledge cannot in and of itself define the situation, the "facts of the situation" are an important constraint on the range of possible interpretations. . . . Given this interpretive dimension, science loses its privileged claim as superior knowledge. Empirical science need not fold up shop, but in a practical field like public policy, it has to establish a new relationship to the other relevant discourses that bear on policy judgments. . . . Instead of questioning the citizen's ability to participate, we must ask, How can we interconnect and coordinate the different but inherently interdependent discourses of citizens and experts? (F. Fischer, 2000, pp. 44–45)

Following F. Fischer's lead, I wish to explore how in settings such as energy and environmental policy debates, the apparent incompatibility of lay and professional perspectives can be overcome and these divergent approaches can be integrated productively. I focus here on one key prerequisite for the interconnection and coordination that Fischer recommends. To counter monolithic technocratic decision making, or better yet, to engage in productive collaboration with technical specialists, members of the public must have reasonable fluency in the language(s) of science. Here, I call this fluency *public expertise*. The ideal form of public expertise is technical competency acquired and used directly by affected citizens. Such competency need not, and cannot, replace the more specialized knowledge of technical or policy professionals, but it can provide members of the public with an adequate foundation for genuine dialogue with these specialists. When this ideal is unreachable, public expertise can also take the form of technical knowledge made available to the public by supportive specialists through consultation, advising, education, or facilitation of citizen-directed research.

The concept of public expertise may seem paradoxical. As Lyotard (1984) emphasizes, contemporary understandings of expertise are closely linked to the principle of specialized knowledge. Gieryn (1995), Taylor (1996), and others have demonstrated that *boundary work* or *demarcation* is a fundamental rhetorical activity for expert communities, serving to maintain their cognitive and political autonomy by preserving distinctions between experts and the public. Yet as F. Fischer (2000) demonstrates, ordinary citizens possess important insights regarding the practical contexts that give meaning to expert discourse. If expertise consists of understanding particular kinds of problems comprehensively, in all their relevant dimensions, then it must incorporate the local knowledge and evaluative contexts that ordinary citizens provide. In this respect, members of the public are experts, too, with their own forms of special knowledge. Conversely, when professional experts contribute to public dialogue on energy and environmental issues, or on other issues with technical dimensions, they do so as interested members of a larger community. Expertise is ultimately a public resource, and one emerging challenge for professionals is to increase the degree to which their knowledge is publicly available. Thus the boundaries between expert knowledge and public knowledge, or between experts and the public, are growing increasingly unclear.

I look more closely at this convergence of expert and public knowledge, outlining some implications for the relationships between professional experts and the public. My discussion draws on the literatures of conflict communication and environmental communication, as well as on Fischer's analysis, which is strongly influenced by communication theory and by contemporary social studies of science. Additionally, I incorporate a number of insights from scholars and practitioners who have examined public participation in environmental decision making. My synthesis of these materials is informed by my own experiences as a citizen advisory board member and as participant-researcher in the area of environmental communication.[1] Drawing on these sources, I argue that whereas increased public participation is necessary in the discourse of energy and environmental issues, public participation is constrained by the traditional boundaries between experts and the public. In particular, technical competency constitutes a formidable practical and symbolic barrier to increased citizen participation. Re-conceptualizing expertise as a public resource, and the relationship between experts and the public as a civic dialogue, is essential to overcoming that barrier and to improving the quality of energy and environmental decisions.

A RATIONALE FOR PUBLIC PARTICIPATION

W. R. Fisher (1987) introduces his model of public moral argument by examining the critical reception of a now classic ecological manifesto, Jonathan Schell's *The Fate of the Earth* (1982). Fisher argues that despite its fundamental rationality, Schell's call for the abolition of nuclear weapons was dismissed by numerous technical experts as unrealistic and impractical. In W. R. Fisher's analysis, "public moral argument was thus overwhelmed by privileged argument." As part of this process, "technical argument . . . denuded it of morality altogether, making the dispute one for 'experts' alone to consider" (Fisher, 1987, p. 71). Although this example focuses on a widely read book written to provoke debate at the broadest public level, the same logic governs most discourse on energy and environmental issues, from local community disputes to debates on national policy. Paradoxically, although these are clearly public questions, the most widely persuasive arguments are those based on special expertise. In W. R. Fisher's analysis,

> [*P*]*ublic* moral argument is a form of controversy that inherently crosses professional fields. It is not contained, in the way that legal, scientific, or theological arguments are, by their subject matter, particular conceptions of argumentative competence, and well-recognized rules of advocacy. Because this is so and because its realm is public-social knowledge, *public* moral argument naturally invites participation by field experts and tends to become dominated by the rational superiority of their arguments. *Public* moral argument, which is

> oriented toward what ought to be, is often undermined by the "truth" that prevails at the moment. The presence of "experts" in *public* moral arguments makes it difficult, if not impossible, for the public of "untrained" thinkers to win an argument or even to judge arguments well. (Fisher, 1987, p. 71; emphasis in original)

Fisher attributes this discursive collapse to an excessive reliance on what he calls the "rational-world paradigm." His comments resonate with Habermas' critique of "instrumental-technical rationality" (Habermas, 1984), as well as with Lyotard's concerns about the contemporary obsession with "performativity" and the corresponding loss of alternative legitimating narratives (Lyotard, 1984). Within these "norms of rhetorical culture" (Farrell, 1993),

> Only experts can argue with experts, and their arguments—although public—cannot be rationally questioned by nonexperts. . . . In the presence of experts—those best-qualified to argue, according to the rational-world paradigm—the public has no compelling reason to believe one expert over the other. Nonexperts cannot be judges; they become spectators whose choice becomes only a nonrational choice between actors on a stage. (Fisher, 1987, p. 72)

In Pearce and Littlejohn's terms, such a discourse is rhetorically attenuated, and as a result of the transformation of nonexperts into spectators it is also morally attenuated. Fisher offers an alternative vision for public moral argument, however—a "narrative paradigm" in which all interested parties participate equally:

> Within the narrative paradigm's perspective . . . the experts' stories are not at all beyond analysis by the layperson. The lay audience can test the stories for coherence and fidelity. The lay audience is not perceived as a group of observers, but as active, irrepressible participants in the meaning-formation of the stories that any and all storytellers tell in discourses about nuclear weapons or any other issue that impinges on how people are to be conceived and treated in their ordinary lives. (p. 72)

A transition from the rational-world paradigm to the narrative paradigm entails changes in the roles of both experts and members of the public, as well as in how the very concept of knowledge is understood. Expert knowledge becomes one form of narrative knowledge, indispensable in areas such as energy and environmental policy but by no means self-sufficient. Accordingly, experts retain authority as custodians of one highly valuable, but partial and positioned, form of knowledge. The proper role of the expert is to make this knowledge available within a larger discursive field, as an active participant in the community's production of meaning rather than as a unitary source of meaning:

> From the narrative perspective, the proper role of an expert in public moral argument is that of a counselor, which is, as Walter Benjamin notes, the true function of a storyteller. . . . His or her contribution to public dialogue is to impart knowledge, like a teacher, or wisdom, like a sage. It is not to pronounce a story that ends all storytelling. An expert assumes the role of public counselor whenever he or she crosses the boundary of technical knowledge into the territory of life as it ought to be lived. (p. 73)

Embracing such a role requires a substantial adaptation by the technical expert, who moves from a position of privilege to one of equal partnership. No less drastic, however, is the change in the role of the ordinary citizen. As active and equal participants in policy discourse with an undeniable technical dimension, members of the public must be capable of listening to, evaluating, and contributing to technical conversations. Here Frank Fischer's analysis complements that of Walter Fisher. F. Fischer (2000) maintains that public participation is valuable in three ways: it cultivates democratic politics and thereby counters the trend toward technocratic control and individual alienation; it provides legitimation for particular decisions as well as for public institutions; and it enhances the relevance and validity of technical analysis at all levels of decision making. It is the third of these motives that Fischer examines most closely and most effectively, focusing on citizens' abilities to contribute as well as on the usefulness of their contributions in particular policy decisions.

Drawing on examples from Europe, Asia, and the United States, Fischer argues that "many citizens are much more capable of grappling with complex technical and normative issues than the conventional wisdom would have us believe" (p. 260). His examples include cases of "popular epidemiology" (Brown & Mikkelsen, 1990), successful citizen panels and juries (Crosby, 1995), the widely acclaimed European "consensus conferences" (Joss & Durant, 1995), and the "participatory research" movement (Cancian & Armstead, 1992; Laird, 1993; Reason, 1994). His goal is to identify viable mechanisms for substantive public participation in environmental decisions as a step toward what Dryzek (1990, 1996) calls "discursive democracy." As a leader in the "argumentative turn" or "communicative turn" in the public policy discipline (Fischer & Forester, 1993; Healey, 1993), Fischer maintains that successful policy flows from a broad and comprehensive dialogue in which citizens articulate, interrogate, and transform each other's perspectives. Within this dialogue, the local knowledge of ordinary citizens and the abstract knowledge of technical experts interact synergistically to provide more complete analyses and more effective decisions.

PUBLIC EXPERTISE AND THE "PARTICIPATION GAP"

Fischer provides considerable evidence that citizens are capable of direct participation, and that such participation is essential both to democratic politics and to successful decision making. Nevertheless, reviews of existing approaches to public participation, especially as practiced in the United States, suggest that these often fall far short of Fischer's ideal (Chess & Purcell, 1999; Fiorino, 1990, 1996; Laird, 1993; McComas, 2001; Rowe & Frewer, 2000). Fiorino (1996) has described this deficit as a *participation gap* in environmental policymaking.

Acknowledging the limits of technocratic decision making and responding to calls for more openness, U.S. federal and state agencies have developed a variety of public involvement methods including public meetings and hearings, citizen advisory boards, and citizen panels and juries. Meanwhile, ballot initiatives and referenda have become increasingly influential in state and local environmental politics. These approaches are certainly valuable and represent progress toward a more dialogic model. Nevertheless, they all have significant limitations related to real or perceived limits on the abilities of ordinary citizens to participate in technical discourse.

Public meetings, widely used by the U.S. Department of Energy and the U.S. Environmental Protection Agency, are attempts at direct democracy but typically provide only limited opportunities for citizen engagement with issues and decision makers. In many cases, this engagement is brief and comes only after the issues have been framed and a narrow menu of policy choices has been developed. In the worst cases, meetings provide "hollow participation in which citizens merely make noise in some political ritual" rather than "real influence over outcomes" (Laird, 1993, p. 348). As Fiorino (1996) notes,

> In most public hearings the agency defines the agenda and establishes the format. The hearing itself provides limited time for citizens to understand the technical or policy issues and to take a substantive part in the discussion. Indeed, the reliance on public hearings as a mainstay of public participation is one of the weaknesses of the administrative process in the United States, in part because of the unequal relationship of citizens to government officials. . . . Public hearings typically do not give citizens a share in decision making. Although they provide mechanisms for public views to come to the attention of administrators, they do not directly engage citizens in the process of making policy choices or cede to citizens any control over the decision process itself. (p. 202)

Panels, juries, and advisory boards provide small numbers of citizens with opportunities for more extended engagement, but these more highly involved individuals do not necessarily represent the larger public that remains distanced from the issues. Appointments to such bodies often emphasize interest group politics over direct democracy (Fiorino, 1996; Laird, 1993; Williams & Matheny, 1995). In addition, public-interest representatives who serve in these bodies for extended time periods run the risk of "going native" by becoming specialized experts themselves. As these individuals increasingly identify with their formal roles, they may lose contact with the communities and values that they are presumed to represent. In Fischer's view,

> [I]nterest group politics are not to be misconstrued with citizen involvement in the sense at issue here. Although they speak in the name of large numbers of people, such groups are typically run by a small group of people at the top of their organizations. Indeed, interest group politics has seldom proven to be participatory democracy in action. . . . Many grassroots environmentalists in the United States, especially those identified with the environmental justice movement, strongly complain that the big environmental Washington-oriented organizations have lost touch with the local citizenry. Having become caught up in so-called "Beltway" politics, such organizations increasingly represent their followers only on paper. (F. Fischer, 2000, pp. 33–34)

Similarly, ballot initiatives are often developed by interest groups and brought to the polls based on popular support of their perceived or claimed purposes, rather than on close scrutiny and broad substantive participation. Understanding the full implications of a ballot measure often requires a substantial familiarity with the underlying technical and policy issues; as a result, voters rely heavily on opinion leaders and media representations to interpret the meanings of ballot measures. Furthermore, in this mode of action "the influence that any one person can have is small. One of the many weaknesses of initiatives is that they force people to make dichotomous choices, which offers them a very limited kind of decision authority" (Fiorino, 1996, p. 202).

The participation gap manifested in these practices has multiple roots. One of these is the sheer complexity, pace, and breadth of contemporary society, which seems to offer individuals no alternative to reliance on specialists in myriad technical fields (Beck, 1992; Giddens, 1990). Apathy, political alienation, and the many distractions of the consumer society are additional and related factors, although democracy theorists argue that these are products of a thin political culture as much as they are causes of it (Barber, 1984). As Williams and Matheny (1995) have pointed out, prevailing notions of liberal politics offer few conceptual alternatives to a dichotomous choice between technocratic or "managerial" approaches and "pluralist" or interest group models. Both of these paradigms assume—whether correctly or incorrectly—that ordinary citizens lack the competencies required for deep engagement with policy issues. The managerial approach, objectivist in its premises, seeks to solve this problem by delegating decision making to experts who can use objective analytical methods to identify optimum solutions. The pluralist model, relativist in its premises, seeks to do so by facilitating competition and compromise among affected interest groups. Neither approach allows for the broad public dialogue that would characterize the communitarian alternative suggested by Williams and Matheny, a form of "strong democracy" (Barber, 1984) in which ordinary citizens participate directly and have substantial influence.

In this practical and discursive context, F. Fischer (2000) acknowledges that "participation is a challenging and often frustrating endeavor . . . collective citizen participation is not something that just happens. It has to be organized, facilitated, and even nurtured" (p. 260). In support of that goal Fischer and others (Kleinmann, 2000; Sclove, 1995; Williams & Matheny, 1995) seek to enlarge the repertoire of available participation methods, drawing in part on Western European models. These scholars are primarily concerned with institutionalizing new approaches that allow for participation by greater numbers of citizens, over longer periods of time, in greater depth, and with increased authority. However, although their suggestions focus on institutional arrangements that would bring new voices into conversations about energy and environmental issues, they do not address important questions regarding how those conversations are conducted. Technical expertise is one prerequisite for "communicative competence" (Habermas, 1984) in those conversations.

Without adequate technical knowledge, ordinary citizens find it difficult to participate in existing venues and are not well prepared for the more intensive approaches that have been suggested. Unless they perceive that their personal interests are clearly and directly affected, few people choose to enter the intimidating, alienating, and frustrating conversations that surround energy and environmental issues. At the same time, policymakers and technical specialists tend to view public participation as uninformed, irrational, or irrelevant when it diverges from the rules of the prevailing rhetorical culture. Surely, one point of public participation is to enlarge that system of rules, but this is a process that must proceed incrementally. Viewed in dialectical terms, public participation must engage the prevailing discourse to transform it. Viewed in structurational terms (Giddens, 1984), the technical dimensions of energy and environmental discourse are resources at the same time as they are constraints. Thus, one important tool for effective citizen participation is technical competency, and in most cases citizens must possess some basic level of technical knowledge to enter the conversation at all.

CULTIVATING PUBLIC EXPERTISE

Ordinary citizens need not acquire the same depth of technical knowledge as specialists; indeed, their doing so would entail becoming specialists themselves and would impair their status as representative members of the public. In this regard, Fiorino (1996) emphasizes the importance of "direct participation of amateurs . . . as citizens engaged in governance . . . rather than professionals doing a job" (p. 200). Nevertheless, to succeed as amateur, expert members of the public must possess basic technical competencies. At the most general level, these include a working vocabulary of scientific terms and concepts, and an overall understanding of how technical reasoning operates. Understood as technical literacy, these competencies are among the goals of formal education, but the rapid changes characteristic of contemporary society require that they be continually refined. Basic technical knowledge of this sort enables citizens to follow evolving policy issues, increases the likelihood that they will take an active interest in these issues, and prepares them for more successful involvement with particular issues.

In the discourses of particular policy issues, citizens encounter more specialized terms and concepts, jargon, acronyms, and methodological questions, as well as related legal, administrative, and regulatory details. Effective public participation requires critical engagement with all these components of discourse. Here, technical specialists must not only be providers of information and analysis; they must also educate people regarding the interpretation and use of those technical findings:

> [P]articipation must be meaningful. Part of that requirement is that citizens be educated about the issues at hand and what they can do to influence policy decisions. . . . In part, this criterion means that relevant information must be provided to citizens, but information is not enough. Inundating the people with mountains of raw data is not a democratic exercise. Rather, citizens must be given information and analysis that are genuinely educative. Citizen *understanding* must improve. (Laird, 1993, pp. 347–348)

As Laird recognizes, information, facts, and data are now more available than ever before (Kinsella, 2001; U.S. Department of Energy, 1995b). However, to make use of these citizens, one must understand them in their technical and policy contexts:

> [I]t is not enough that participants simply acquire new facts. They must begin, at some level, to be able to analyze the problem at hand. At the simplest level, this means understanding the different interpretations that one can draw from the facts and trying to think about ways to choose among those interpretations. At a more sophisticated level, it means beginning to learn how and when to challenge the validity of the asserted facts, where new data would be useful, and how the kinds of policy questions being asked influence the type of data they seek. Perhaps more important, analyzing a problem means being able to challenge the formulation of the problem itself, that is, for people to decide for themselves what the most important questions are. (Laird, 1993, pp. 353–354)

Here Laird is calling for a type of citizen competency that extends beyond basic technical literacy. To participate effectively, and to integrate the results of technical analysis with their own local knowledge and evaluative criteria, citizens also require a broader and more critical understanding of the rhetoric and sociology of technical discourse. F. Fischer (2000) and others have assembled substantial evidence that this kind of citizen understanding is possible, but to achieve this goal citizens, technical specialists, and policy specialists must collaborate closely. F. Fischer, like Laird, emphasizes the educational role that specialists must play as part of that collaboration:

> For the professional disciplines, this poses the challenge of rethinking the professional–client relationship. In such cases, the professional must learn to adopt more cooperative and facilitative interactions with the citizen–client. This, as we have seen, shifts the professional role from that of authoritative advisor to facilitator of client discourse. For this practice, professionals must develop a quite different set of skills. Rather than just to offer packaged solutions, the facilitator's task is to conceptualize and present policy alternatives and arguments for public deliberation. Beyond a competent grasp of empirical–analytic skills, he or she requires as well the ability to effectively share and convey information to the larger public. In this sense, the analyst is as much an educator as a substantive policy expert. (F. Fischer, 2000, p. 261)

As consultants, facilitators, educators, or sages, specialists place their expertise into a larger civic conversation where it becomes a resource for public decision making. For that resource to be useful, it must include not only technical data and analyses but also guidance on how to understand and evaluate those technical products. These contributions can be viewed as the local knowledge of specialists, that is, their specific contributions to the larger dialogue. Viewed this way, specialists and nonspecialists assume parallel and complementary roles in the production of public expertise.

CONCLUSION: PUBLIC EXPERTS AND AN EXPERT PUBLIC

A clear distinction between experts and the public has been one aspect of the technocratic approach to energy and environmental policy. As this technocratic model has proved inadequate, mechanisms for public participation have evolved, but these fall short of providing broad, direct, and substantive citizen engagement. Enhancing public participation is important for preserving and strengthening democratic politics, for enhancing the legitimacy of public decisions, and for incorporating the knowledge and values of affected parties into those decisions. Nevertheless, the expert–public distinction continues to pose a practical and symbolic barrier to participatory decision making. Viewing expertise in broader terms, as a public resource created through broad dialogue, is one way of reducing that barrier.

Ordinary citizens provide their own forms of expert knowledge when they contribute their local perspectives and values to policy decisions. Similarly, the contributions of specialists can be seen as one kind of local knowledge in the sense that they are situated and partial perspectives that have specific relevance within public conversations. Thus the distinction between experts and nonexperts is exaggerated and largely artificial. With appropriate support from specialists, members of the public can develop technical competencies adequate for substantive participation. By making their own expertise more public, specialists can, in turn, benefit more fully from the complementary knowledge that nonspecialist citizens provide.

NOTE

1. Since April 2000, I have served on a citizen board that advises the U.S. Department of Energy, the U.S. Environmental Protection Agency, and the Washington State Department of Ecology on the cleanup of the Hanford Reservation, the most polluted site within the nation's former nuclear weapons complex. In that capacity I have closely observed how that board makes use of public expertise. As a participant in technical education workshops offered by the Institute for Energy and Environmental Research (IEER), and as a member of the Institute's education committee, I have examined one organization's efforts to develop public expertise through education, training, and collaboration with citizen activists. Although I do not discuss these experiences directly in this chapter, they have provided an important interpretive framework for my comments.

Part Two

Evaluating Mechanisms for Public Participation in Environmental Decision Making

Chapter 5

Decide, Announce, Defend: Turning the NEPA Process into an Advocacy Tool Rather than a Decision-Making Tool

Judith Hendry

Since its signing into law by President Nixon on January 1, 1970, the National Environmental Policy Act (NEPA) has done much to merit Henry "Scoop" Jackson's characterization as the most important and far-reaching environmental and conservation measure ever enacted by Congress. In a piece of legislation barely three pages long, NEPA provided the framework for incorporating human values and place-based identity into the decision-making process, thus giving a voice to those who must bear the economic, social, and environmental consequences of government policy and land-use decisions. For the first time, government agencies were required to prepare an environmental analysis that integrated input from the stakeholders, state and local governments, and Indian tribes when considering a major land-use proposal.

One of NEPA's defining innovations was its emphasis on including the public as consultants in federal agency decisions. Prior to NEPA, the public had little opportunity to engage in dialogue concerning the social, economic, and environmental costs and benefits of an agency's land-use decision. Nor was there much recourse to challenge these decisions once they were made. The purpose of giving the public a voice was to help ensure that the agency was better informed about the consequences of its decisions. Public input aids in gathering baseline data, helps to identify key issues that may not otherwise surface, and serves to build trust among the agency and the stakeholders (Council on Environmental Quality [CEQ], 1997).

Although the intent of NEPA is laudable, its implementation is often cumbersome, costly, and regarded as a compliance hoop through which agencies must jump before going ahead with a planned action (Council on Environmental Quality, 1997). This "decide, announce, defend" strategy (Yosie & Herbst, 1998) undermines the intent of NEPA by viewing public participation as an end rather than as a means to decision making and the NEPA process as merely an instrument to validate a priori decisions. This paper explores what, I contend, is just such a case in which the NEPA process was used as a tool to advocate a controversial decision rather than as a decision-making tool.

Previous research within the field of environmental communication has addressed the inadequacies inherent within the NEPA process and made recommendations regarding how agencies can better incorporate the tenets of NEPA into their public input processes and decision making. Graham (1997) suggests an alternative model of public participation from a social communication perspective that emphasizes openness and shared responsibilities among agencies and stakeholders. Walker and Daniels (1997) compare various collaborative approaches that can be adopted to increase the effectiveness of the often confrontational public input processes. In a similar vein, Waddell (1996) discusses what he terms a *social constructionist* model of public participation that eschews the traditionally constructed science-versus-values dichotomy. This model instead calls for an interactive exchange that engages the values, beliefs, and emotions of all stakeholders to socially construct policy decisions.

Although these alternative approaches hold promise for improved implementation of public input into agency decision-making processes, they presuppose a willingness on the part of the government agency to go beyond the minimum requirements outlined by NEPA—to establish trust with stakeholders and relinquish power to allow for joint decision making. It is unfortunate that in over 30 years of practice there is little evidence to suggest that federal agencies have been willing or able to successfully implement these kinds of collaborative processes.

An extensive study conducted in 1997 by the Council on Environmental Quality (CEQ), the agency responsible for overseeing the implementation of NEPA, found a number of problems with the public scoping and input processes. The following is a list of some of the participants' frequently stated observations and concerns:

- Citizens felt that they were treated as adversaries rather than as welcome participants.
- Citizens felt that they were invited too late to discuss an already well-developed project.
- Citizens felt that their input was not reflected in changes to the proposal.
- Citizens felt overwhelmed by the resources available to agencies and proponents of the action or project.

The study concluded that "substantial opportunities exist to improve the effectiveness and efficiency of the NEPA process" (p. 35). However, before improvements can be made, it is necessary to have a clear understanding of how the process can be undermined. This case study examines how the NEPA process was used to support an a priori decision and discusses the specific strategies employed in turning a decision-making tool into an advocacy tool.

The implementation of public input processes and the ways that this input is then used in decision making is defined in the NEPA Handbook. This document provides "instructions for complying with the CEQ Regulations for Implementing the Procedural Provisions of the National Environmental Policy Act of 1969" (NEPA Handbook, 1988, p. i). Because this study pertains to land-use decision involving the Bureau of Land Management (BLM), the following overview is based on the BLM's NEPA Handbook.

A BRIEF DESCRIPTION OF THE NEPA PROCESS AND TERMINOLOGY

NEPA requires federal agencies to prepare a detailed statement of the environmental impacts of a proposed action. An *Environmental Assessment* (EA) is a concise public document that examines the impacts on the quality of the human environment. The CEQ recommends that the EA document should be no longer than 10 to 15 pages exclusive of appendices. If, on completion of the EA, it is determined that the proposed action will result in no significant adverse impacts, the agency will then issue a *Finding of No Significant Impact* (FONSI) and the proposed action can be approved.

Another version of the FONSI is the *Mitigated Alternative* often referred to as a *Mitigated FONSI*. The CEQ report (1997) clearly explains the concept of the mitigated FONSI:

> While preparing EAs, agencies often discover impacts that are "significant," which would require preparation of an EIS. Agencies may then propose measures to mitigate those environmental effects. If an agency finds that such mitigation will prevent a project from having significant impacts on the environment, the agency can then conclude the NEPA process by issuing a FONSI rather than preparing an EIS. The result is a "mitigated FONSI." (p. 20)

If, on the other hand, the EA concludes that the impacts may be significant, the action cannot be approved unless it is put through a more detailed and stringent analysis called an *Environmental Impact Statement* (EIS). If, at the outset of the decision-making process, the agency believes that the proposed land-use action will have significant environmental impacts or that the action will be highly controversial, it may dispense with the EA and go directly into the more rigorous EIS process. Because the EIS is a detailed, long, and costly process, the agency may instead decide on the *No Action Alternative,* which simply means that the agency decides not to approve the proposed action.

Once the EA or EIS is completed, the agency makes a determination of whether or not to approve the proposed action. Affected parties have the right to appeal the agency's decision through an administrative process. In the case of BLM, appeals are decided by the Interior Department's Board of Land Appeals.

The preparation of an EA rather than an EIS is the most common source of conflict and litigation under NEPA (CEQ, 1997). Yet despite the increased risks of challenges, there has been significant increase in the number of EAs and a decrease in the number of EISs prepared each year.

> All signs point to a significant increase in EAs and a decrease in EISs. The annual number of EISs prepared has declined from approximately 2,000 in 1973 to 608 in 1995. By 1993, a CEQ survey of federal agencies estimated that about 50,000 EAs were being prepared annually. The survey also found that five federal agencies—the U.S. Forest Service, the Bureau of Land Management, the Department of Housing and Urban Development, the U.S. Army Corps of Engineers, and the

> Federal Highway Administration—produced more than 80 percent of the EAs. (CEQ, 1997, p. 19)

Likewise, the CEQ reports a significant trend in the use of the mitigated FONSI as was issued in the following case study. The core of NEPA's strength lies in its requirement for rigorous analysis, its mechanism for public involvement, and the ensuing consideration of multiple issues and alternatives. The decreasing number of EISs and the increasing use of mitigated FONSIs led the CEQ to offer the following admonition:

> When the process is viewed merely as a compliance requirement rather than as a tool to improve decision-making, mitigated FONSIs may be used simply to prevent the expense and time of the more in-depth analysis required by an EIS. The result is likely to be less rigorous scientific analysis, little or no public involvement, and consideration of fewer alternatives, all of which are at the very core of NEPA's strengths. (CEQ, 1997, p. 20)

THE COMMUNITY AND THE CONTROVERSY

Lured by spectacular vistas, the rural setting, the beauty of the high desert terrain, and the easy commute from both Albuquerque and Santa Fe, many new residents have been drawn to Placitas, New Mexico. The once sparsely inhabited area surrounding the historic village of Placitas is now enveloped by upscale adobe-style homes high up on the ridges and overlooking the arroyos. The population of this unincorporated area is estimated to be about 4,000 and is expected to triple by the year 2020.

As is the case with many rapidly developing communities, residents are experiencing the growing pains of urban development. One concern of particular relevance to this case study centers around recreational open space. The BLM oversees a 4,000-acre tract of land just to the north and west of the Placitas area that is used by local residents for various recreational purposes including hiking, horseback riding, and target shooting.

The BLM lands also support a wide variety of plants and animals. Las Huertas Creek meanders across the BLM lands on its way to the Rio Grande and serves as an important watershed and recharge area for mountain snow melts. Recent archeological surveys have revealed that the creek has been a lifeline for generations past. There are over 80 recorded archeological sites along a one-mile stretch of the creek, 17 of which are eligible for the National Historic Register (Daniels & Schutt, 1999). Additional concerns for the preservation of these BLM lands revolve around their use for ceremonial purposes by neighboring Indian tribes.

Spurred by concerns for the encroaching development and the increased recreational use of the open space, residents formed Las Placitas Association (LPA) in 1995 "to preserve and protect open space in the Placitas area for recreational, educational, and rural activities and to promote environmental and cultural preservation" (Las Placitas Association, 1999). In the past five years, this nonprofit organization has gathered a great deal of community support for its efforts and has raised over $100,000 for open space preservation.

In 1997, BLM received a proposal from Western Mobile to lease a portion of its Placitas holdings for gravel mining. At the time, Western Mobile was mining on privately owned land adjacent to the BLM lands, and this proposal would expand its operations on to BLM lands. Upon receipt of the proposal, the BLM began the EA process, and Western Mobile contracted a local environmental consulting firm to prepare the EA document. While the contractors were hired by Western Mobile, the BLM had the ultimate say concerning whether the final document was accurate, complete, and unbiased.

In view of the multiple and diverse stakeholders who would be affected by the decision of whether or not to allow this mine to go forward, it is not surprising that the public scoping meeting and the ensuing public response period produced a great deal of input in which numerous concerns were expressed.

Gravel mining is an expressly incompatible neighbor to public recreation and residential areas. In addition to the environmental and aesthetic impact of strip mining 280 acres of the fragile desert terrain, there were many other concerns expressed during the public scoping and response period. These included issues such as the preservation of archeological sites, air quality standards (dust in this arid climate is a big concern), noise pollution, and the impacts of gravel truck traffic as well as an increase in water use. According to the EA, current Placitas gravel mining operations use about 65,000 gallons of water a day for the crushing process and to control the dust. This is approximately equivalent to what 200 homes would use in a day. Other concerns centered around decreased property values for those whose homes would be within sight and hearing of the proposed mine, as well as concerns from the many recreational users of the BLM lands.

Over 500 people opposed to the proposal responded to the BLM through letters, petitions, or input at the two public meetings. Yet despite the tremendous public opposition, a mitigated FONSI was issued, allowing the mine to go forward with minimal mitigation measures added to the original proposal. Shortly after the announcement, Las Placitas Association, two other Placitas residents, and the City of Albuquerque filed appeals with the Interior Board of Land Appeals to reverse the decision. As of the writing of this paper, one of the appeals has been settled and the other three are still pending.

THE EA AS AN ARGUMENT FOR AN A PRIORI DECISION

The means by which the Environmental Assessment (EA) document was used to create an argument in favor of the proposed gravel mine and to justify the decision to allow the mine to go forward comprise the focus of this study. The analysis of the EA document revealed four strategies that were used to create an argument in favor of the decision: (1) the exclusion of recommended areas of inquiry, (2) the use of an unpopular alternative as a coercive threat, (3) conceding superficial accommodations to community concerns, and (4) framing the action as a quantifiable, utilitarian public good. These strategies, whether intentionally or inadvertently employed, should be recognized as red flags that can undermine the NEPA process and its value as a strategic planning tool.

THE EXCLUSION OF RECOMMENDED TOPICS OF INQUIRY

The recommended format for the EA, as outlined in the NEPA Handbook, calls for the inquiry into and discussion of specifically identified topics. Among the topics to be included is one labeled "Conformity to Statutes, Regulations, or Other Plans." This section of the EA is to address "whether the proposed action [the gravel mine] is consistent with other Federal agency, State, or local plans and programs" (NEPA Handbook, 1988, Section H-1790, p. 1). Although federal lands are not subject to local zoning restrictions, the environmental assessment calls for an examination of the proposed plan's conformity with local plans.

The private lands immediately adjoining the BLM holdings on three sides are zoned residential (by Sandoval County) and are rapidly developing residential subdivisions. The already existing gravel mines on nearby private lands are allowed to continue operating under nonconforming permits until their leases or their gravel reserves expire, but they are not allowed to expand their operations. In view of these local zoning restrictions, it is unlikely that a new mine on the adjoining BLM lands would be deemed "consistent." Yet, this was not discussed in the EA.

The EA followed the recommended format with, however, one conspicuous modification. The section that should have been entitled "Conformity to Statues, Regulations, or Other Plans" was omitted and replaced with a section entitled "The National Environmental Policy Act Process." No mention was made, here or elsewhere, of the proposed mine's inconsistency with local planning, despite the fact that this kind of heavy industrial use is incompatible with residential zoning.

The omission of this section of the EA could be construed as an attempt to distract the reader's attention from a rhetorically deleterious discussion and divert attention to a subject more congenial to the advocacy position that the document supports. Crable (1990) likens this kind of rhetorical strategy to magic in the form of sleight of hand:

> The magic of magicians is based primarily on "misdirection" and "distraction." As the audience of magicians, we look where we have been directed and see what we have been told to see. The magic is not that we see things "appear" and "disappear," the magic is that we fail to see what is really occurring. (p. 123)

As with the sleight of hand, the fact that a section of the EA was omitted and replaced by another section would not be readily apparent to community members unfamiliar with the NEPA Handbook.

What was discussed in the section that was inserted in place of this drew considerable attention from community stakeholders. In fact, the information presented to the reader in this section served to divide the community and pit neighbor against neighbor and, for many community members, effectively chilled opposition to the proposed mine. This section described what was viewed by many as a coercive threat in the form of an unpopular alternative that would be implemented by the gravel company if their BLM lease were not approved.

UNPOPULAR ALTERNATIVE AS COERCIVE THREAT

The section of the EA that was inserted in place of the section called for in the EA format explains to the reader that NEPA requires an assessment of the impacts of the No Action Alternative. In other words, it explains that

BLM must not only assess the impacts of the proposed mine, it must also assess the impacts of not allowing Western Mobile to mine the BLM land. According to the information presented to the reader in this section and again later in the document, if the No Action Alternative were to be accepted and they were not allowed to mine on BLM land, Western Mobile would have to increase its operations at its already existing mine (referred to as the Placitas Pit). The Placitas Pit is currently operating on nearby privately owned land under a nonconforming permit. The expanded operations would, according to the EA, "require the installation of a second crusher, and additional heavy equipment. . . . With increased production from the Placitas Pit, under this alternative, night mining could become necessary, increasing the use of lighting, and increasing noise impacts" (U.S. Bureau of Land Management [US BLM], 1998, p. 14).

A number of residents own property adjoining the boundaries of the Placitas Pit. Several of these residents submitted letters in strong opposition to the proposed mine during the initial scoping period. However, when faced with the threat of increased activity at the neighboring pit, they began organizing support among their neighbors for the proposed mine. Western Mobile sweetened the pot by throwing in an additional no-mining buffer between the boundaries of the Placitas pit lease and adjoining private properties if, and only if, its new mine was approved. In a letter to residents of the subdivision, the president of Western Mobile stated, "Pending the positive outcome of the BLM's decision [to approve their application for a mining lease], Western Mobile would agree to extend the buffer area . . ." (letter dated April 6, 1998 included in the unpublished Administrative Record, available for public review at the BLM office in Albuquerque, NM).

As a result, BLM received 25 letters in support of the proposed mine from neighbors who live closest to the currently operating Placitas mine—many of these letters from the same people who had initially written letters in strong opposition to the proposed mine. The threat of the No Action Alternative had, to say the least, a chilling effect on many community members' willingness to oppose the mine. But perhaps more unfortunate was that the No Action Alternative pitted neighbor against neighbor. Those who owned property close to the proposed mine site felt that they had been "sold out" by those who lived closest to the existing mine.

Although the threat of increased mining activity at the Placitas Pit was twice clearly spelled out, the EA failed to explore whether or not increased mining would be permitted by zoning and other permit restrictions under which the Placitas mine currently operates. The burden of discovery was left to opponents of the mine, despite the fact that community members specifically requested an investigation of these restrictions. Investigations undertaken by community members found that the Placitas Pit was, in fact, restricted from expanding its operations on several counts. First, expanded activity would increase dust and vehicle emissions and the mine was currently operating at 96 percent of the allowable New Mexico Ambient Air Quality Standards for Total Suspended Particulates. Second, the addition of a second crusher would demand increased water use as large amounts of water are required in the crushing process. Western Mobile was currently using nearly all of its water allotment. Finally, it was operating under a nonconforming zoning permit which expressly prohibited the expansion of its activities.

Upon learning of these restrictions and realizing that expanded activities at the Placitas pit would most likely be prohibited, Las Placitas Association filed an appeal with the Interior Board of Land Appeals (IBLA). Las Placitas Association President, Carol Parker, explained, "It was the loss of trust in BLM that finally made us decide to spend our hard-earned money to launch a legal challenge" (personal interview, February 22, 1999).

Trust between the BLM and community members was further weakened by the approval of the mine with what many viewed as minimal concessions toward mitigating the negative impacts of the mine that were spelled out in the mitigated FONSI. The concessions were viewed as superficial accommodations in the face of their concerns.

SUPERFICIAL ACCOMMODATIONS TO COMMUNITY CONCERNS

The final decision of the BLM was a mitigated FONSI. As explained previously, this means that the mine, as originally proposed, would have significant impacts but that changes to the plan would mitigate the impacts to no longer constitute significant impacts. The mitigation measures called for increasing the mine's distance from Las Huertas Creek from 75 to 150 feet; creating a 75-foot buffer between the mining boundaries and residential properties; further limiting the hours of mining operations to 11 hours a day as opposed to a 12-hour day as originally proposed; the relocation of several rare species of cacti; fencing an archeological site; and setting the back-up alarms on operating equipment to the minimum allowable noise-level settings.

BLM further promised in the mitigated FONSI that "[s]hould unforeseen impacts occur to Las Huertas Creek due to Western Mobile's mining operations, Western Mobile would respond, in a timely fashion, with its best efforts to mitigate the problem" (US BLM, 1998, p. 14). This assurance, although elegant in its abstraction, offers little compensation for those concerned with the mine's potential impact on this valuable groundwater recharge and watershed area—concerns expressed by many during the EA scoping and response period. The following statement from Del Agua Institute, an organization of hydrologists and geologists, summarizes these concerns:

> The landscape and water regime of the high desert near the Sandia Mountains are dynamic, like many desert regions of the Southwest. As the information and data indicate, the proposal to mine aggregate will significantly impact ground and surface water flow. The most obvious direct hydrologic impact is degradation of the existing riparian zone along the drainage. A less obvious, but vitally important potential long-term impact is decreased ground water and surface recharge into the Rio Grande basin, an already over-allocated resource on which the prosperity of the Middle Rio Grande basin is completely dependent. (letter written by Dr. Rebecca M. Summer dated September 19, 1997 and attached to the EA)

Over 500 community members responded in force by giving input during the public scoping meetings and in letters to the BLM. Strong concerns for the potential as well as the inevitable impacts of strip mining 275 acres of fragile desert terrain in close proximity to a critical water recharge area and to a rapidly expanding residential area were expressed. Yet the mitigation measures conceded by the BLM in the mitigated FONSI were comparatively unsubstantial.

The consequences cannot be made insignificant simply by making a few small changes to the plan. Such things as lowering the volume of the backup alarms on equipment and restricting mining to one less hour per day appear to be superficial accommodations in view of the substantial and numerous concerns expressed by community members.

I would even suggest that a mitigated FONSI is, in and of itself, an oxymoron. If the impacts of a proposed action are significant and therefore require mitigation, how can a "finding of no significant impact" be justified? The mitigated FONSI says, in effect, "we are implementing measures to mitigate the significant impacts of which there are none." Nevertheless, if an agency claims that the significant impacts can be mitigated to significance, the mitigation measures would need to be substantive.

Framing the Proposed Action as Quantifiable, Utilitarian Public Good

Snow and Benford (1992) define framing as "an interpretive schemata [*sic*] that simplifies and condenses the 'world out there' by selectively punctuating and encoding objects, situations, events, experiences, sequences, and actions" (p. 137). It is not surprising that the interpretive schemata of those opposed to the proposed mine focus on such place-based impacts as the destruction of the biotic community, the loss of recreational open space, and the negative impacts associated with heavy industry in close proximity to residential areas. The EA, on the other hand, frames the discourse in terms of the quantifiable "public good" and invokes the "utilitarian principle of the 'greatest good for the greatest number'" (Oravec, 1984). The one-page introduction to the EA discusses the public's need for gravel:

> Western Mobile believes that the maintenance and expansion of the road system in the counties and State are necessary to facilitate economic transportation of goods, services, and people. Additionally, the expansion of the population with the attendant expansion of new homes and businesses in the two counties increases the need for concrete and asphalt related building materials. (US BLM, 1998, p. 1)

Framing the proposed action in terms of "public interest" or "public good" infers that there is a public need for the mine. Although it is hard to argue with the public's need for gravel, the unstated, but implied, premise that the public will be harmed if gravel is not extracted from this particular portion of land is questionable. The fact that there are numerous gravel companies operating within a thirty-mile radius of the proposed site leads one to suspect that the gravel needs would be willingly met by Western Mobile's competitors.

The availability of gravel resources for a growing metropolitan area is indeed an important question and one that should have been explored (but was not) in the Environmental Assessment. However, the bigger framing issue is not whether the public does or does not have enough gravel reserves to meet their needs. The bigger framing issue has to do with the definition of "public interest" that is implied within the frame. Defining "public interest" in terms of the public's need for gravel tacitly insinuates that the not-in-my-backyard, self-serving motivations of those who oppose the mine are in conflict with the altruistic motives of the gravel company who is mining on behalf of the "greater public good." The opening discussion of the EA could have just as legitimately (and as questionably) defined "public interest" by framing the discourse as the commercial motives of the multinational corporation in conflict with the altruistic preservation motives of the community.

Bruner & Oeschlaeger (1994) point out that "whoever defines the terms of the public debate determines its outcomes. If environmental issues are conceptualized, for example, in terms of 'owls versus people,' then owls (and the habitat that sustains them) do not have much of a future" (p. 218). In this case, defining "public good" in terms of the utilitarian need for gravel offers an easily identifiable and quantifiable public good. In contrast, the "public good" framed in terms of needs expressed by the local community—to preserve the natural surroundings, to protect the historical and cultural heritage of the area, to maintain the quality of life, to provide recreational open space, and so forth—are far more abstract and difficult to quantify.

One of the principal purposes of NEPA is to provide a forum in which "the unquantified environmental amenities and values may be given appropriate consideration along with economic and technical considerations" (NEPA, 1969, Section 102). When the first page of the EA frames the discourse only in terms of the quantifiable need for the extraction of resources, the ensuing document will most likely be positioned firmly within that utilitarian frame of quantifiable resource needs.

The fact that the need for gravel is punctuated in the first page of the EA is not surprising considering that the document was prepared by contractors who were hired by Western Mobile. This is not an unusual occurrence. The proponents of an action are frequently held responsible for the expense and implementation of the NEPA process. BLM has the authority of final approval of the document and the subsequent decision concerning the land use. Nevertheless, when prepared by, or financed by, the proponent of the action, the document on which a decision will be based runs the risk of presenting an a priori case with built-in biases, however subtle, toward a decision in its own favor. Opponents to the action are subsequently faced with the burden of proof—that is, they are put in a position of having to prove the document wrong, inadequate, or biased. And they must do this without the benefit of the time, resources, or expertise that went into preparing the EA, allowing them little chance for a fair and equal hearing. As such, the writers of the document have a tremendous rhetorical advantage in that they frame the argument from the first to the last page of the EA document.

CONCLUSION

It is in the best interest of both the agency and the stakeholders to demonstrate a good faith effort to incorporate public input into the decision, and NEPA provides the structure and systematic process to accomplish this. When the process is treated as merely an exercise in compliance and when the EA is used as a advocacy tool to support an a priori decision, the decision will inevitably be met with suspicion and confrontation. The resulting appeals and litigation are far more cumbersome, take far longer to resolve, and are far more costly than is an inclusive, fair, comprehensive public input process.

Place-based stakeholder processes inevitably favor industrial interests because industries tend to have disproportionate resources and skills that allow them to use the process in their own favor (Yosie & Herbst, 1998). When the Environmental Assessment is used as a tool for advocating an a priori decision, community members who are negatively impacted by the decision are faced with the burden of proving that the document is inadequate, inaccurate, or biased. And they must do this in a rigidly limited time frame, with limited resources and technical expertise, putting them at a tremendous disadvantage.

As communication scholars we can help to promote improved decision making through critical analyses that uncover the strategies that serve to privilege one voice over another. The list of strategies and prescriptions identified in this chapter is far from inclusive and many more such case studies are required before we can have a comprehensive understanding of the problems that exist within the current practices of implementing NEPA. Nevertheless, this list offers a good starting point from which to begin addressing these problems.

Environmental problems, ". . . in their origins and through their consequences—are thoroughly *social* problems, *problems of the people*, their history, their living conditions, their relation to the world and reality" (Beck, 1992, p. 81). As environmental communication scholars, "It becomes incumbent upon those of us who conduct the research to serve as activists and advocates" (Cantrill, 1998, p. 12). Our theories and research can advance our knowledge of the role of public input in the decision-making process and offer ways to improve and facilitate that process. In so doing we can play a vital role in safeguarding the process that gives voice to the problems of the people who must ultimately contend with the environmental consequences of a government agency's decision.

CHAPTER 6

THE ROADLESS AREAS INITIATIVE AS NATIONAL POLICY: IS PUBLIC PARTICIPATION AN OXYMORON?

GREGG B. WALKER

In 1993, shortly after taking office, President Bill Clinton and Vice President Al Gore heralded the idea of "reinventing government." Vice President Gore became the leader of the Clinton administration's reinvention initiative. The President and Vice President, drawing on Osborne and Gaebler's book, *Reinventing Government*, wanted government institutions to "empower citizens rather than simply serving them" (Osborne & Gaebler, 1992, p. 15).

Six and one-half years later, among the fall colors of the George Washington National Forest in Virginia, President Clinton announced an ambitious idea for America's public lands. The President called for a plan to protect the nearly 40 million acres of roadless areas throughout the United States' National Forest system.

On October 19, six days after the President's announcement, the USDA Forest Service published a notice of intent to implement President Clinton's Roadless Areas Directive. Urged by the Clinton administration to take a fast track approach, the Forest Service subsequently began the initial public involvement process pursuant to requirements of the National Environmental Policy Act (NEPA). As an *Oregonian* article reported on October 14:

> Clinton aides said Wednesday that the U.S. Forest Service will try to speed through hearings and issue regulations that by year's end would

> gain protection for all roadless forest tracts 5,000 acres or larger and perhaps some smaller roadless areas within the agency's 192 million acres nationwide. In addition to a ban on road-building, the agency will consider limits on logging and other development that could cause environmental damage. (Bernton & Hogan, 1999, p. A1)

As part of this fast-track approach, the Forest Service conducted more than 150 public meetings (primarily open houses and a few hearings) during late fall 1999. The fast-track strategy drew criticism for limiting meaningful public participation. An Associated Press article noted that "members of the public had only two days notice of a meeting in Juneau, Alaska, had to draw names from a hat to speak in Portland, Ore., and were told they could not make verbal comments at the Mark Twain National Forest [meeting] in Missouri." Consequently, Senators Bob Smith of New Hampshire and Fran Murkowski of Alaska told Forest Service Chief Michael Dombeck that the process was "replete with . . . fatal flaws" and "to remedy these fatal flaws, the entire . . . process must be started over" (Hughes, 1999, p. A3).

This rapid, centralized, and controversial public participation strategy was initiated by an agency that the Clinton administration's National Partnership for Reinventing Government had designated a "high impact agency." Was the Forest Service's approach an example of reinvention or traditional practice? Former Secretary of Agriculture Dan Glickman has claimed that the Roadless Lands Initiative featured "one of most extensive public-comment processes ever carried out" (Hogan, 1999, p. A7). Does "extensive public comment" translate into meaningful public participation?

More recently, former Forest Service Chief Mike Dombeck expressed similar views. Just after resigning as chief, Mike Dombeck talked about his career with Betsy Marston, editor of *High Country News*. In this March 2001 interview, Marston asked Dombeck to comment on the Roadless Initiative, the policy Dombeck had advised President Clinton to initiate to protect roadless areas throughout the National Forest system. "We've laid out the process," Dombeck noted, "a totally open public process in developing the roads policies: 1.6 million comments, 600 public meetings." "In my entire career," Dombeck added, "this is the most extensive outreach of any policy I've observed" (Marston, 2001, p. 12). During an interview with Steve Curwood of National Public Radio's "Living on Earth" series, Dombeck offered similar comments. "We were very careful to follow the administrative rulemaking process," Dombeck explained. "And 600 public meetings information put out to the public on the Web, mailed out in CD-ROMs and environmental impact statements. In my quarter of a century as a public servant, I don't believe I've seen any issue that has been more thoroughly vetted than this one of roadless areas" (Curwood, 2001). Does the number of public meetings and amount of comment letters received provide sufficient evidence of meaningful public participation?

I have been studying the Roadless Initiative since President Clinton proposed the idea publicly in October 1999. In December 1999, I attended Roadless Initiative "open houses" held by ranger districts of the Siuslaw and Willamette National Forests in western Oregon. During spring 2000, I attended a number of public hearings in western Oregon on the draft environmental impact statement (DEIS) of the Roadless Initiative. I have monitored news coverage of the Roadless Initiative planning process and the related conflict and controversy. The ideas I present here are based on my observations at meetings, my reading of the Draft and Final Environmental Impact Statements, and my interpretation of numerous media accounts and editorials.

PERSPECTIVES ON PUBLIC PARTICIPATION

In environmental and natural-resource policy decision making, public participation and public involvement are often employed as interchangeable terms. At its core, public participation is pre-decisional communication between an agency or organization responsible for a decision and that organization's relevant public community. Public participation as a term is most often applied to a government organization (e.g., federal agency, city government) and the public (Daniels & Walker, 2001).

The National Environmental Policy Act (NEPA), with which the Roadless Initiative planning process must comply, addresses public involvement explicitly. Any federal agency proposing significant environmental actions must conduct an environmental assessment (EA) or an environmental impact statement (EIS). Both require public involvement, although "in practice, the degree of public involvement for an EIS is generally far greater than for an EA" (Eccleston, 1999, p. 70).

The National Environmental Policy Act, passed in 1969, institutionalized public participation as a part of environmental planning, regardless of the environmental context. NEPA did not require collaboration, consensus, or dialogue—only some form of involvement. Since NEPA's inception, agencies have typically interpreted its public participation requirements conventionally, relying principally on hearings and letter-writing comment periods. Figure 6.1 presents the key public involvement features of the National Environmental Policy Act.

Government agencies, specifically those charged with a public participation mandate either under NEPA or similar legislation, have characterized the idea in a variety of ways. For example, in its National Environmental Policy Act Handbook, the U.S. Fish and Wildlife Service (USFWS) notes that "public participation is to be an integral part of the NEPA process. We shall make a reasonable and concerted effort to involve affected Federal agencies, States, government officials and agencies, non-governmental organizations, and the public in the NEPA planning, decision making, and implementation process." The USFWS document emphasizes comment letters or e-mails as the primary method for the public to participate in response to environmental documents. Concerning scoping (identifying issues, concerns, problems), though, the Handbook states that the USFWS "should carefully consider the affected public and provide reasonable advance notice of public meetings and comment due dates to facilitate effective public participation. . . . We should strive to understand the public

Figure 6.1
Specific Directions in NEPA for Involving the Public

Inviting Comments (Section 1503.1)

(a)(4) Request comments from the public, affirmatively soliciting comments from those persons or organizations who may be interested or affected.

(b) An agency may request comments on a final environmental impact statement before the decision is finally made. . . .

Public Involvement (Section 1506.6)

Agencies shall:

(a) Make diligent efforts to involve the public in preparing and implementing their NEPA procedures . . .

(b) Provide public notice of NEPA related hearings, public meetings, and availability of environmental documents . . .

(c) Hold or sponsor public hearings or public meetings whenever appropriate or in accordance with statutory requirements applicable to the agency.

From: *The NEPA Planning Process: A Comprehensive Guide with Emphasis on Efficiency*, by C. H. Eccleston. 1990. New York: John Wiley & Sons.

concerns, accurately record their comments, and allow adequate time for involvement by the affected public" (U.S. Fish and Wildlife Service, 2000). The U.S. Department of Energy's Office of Environmental Management (EM) defines public participation as

> [T]he process by which the views and concerns of the public are identified and incorporated into the Department of Energy's (DOE) decision-making [*sic*]. Furthermore, public participation includes "identifying public concerns and issues; providing information and opportunities for the public to assist [the] DOE in identifying EM-related issues and problems and in formulating and evaluating alternatives; listening to the public; incorporating public concerns and input into decisionmaking [*sic*]; and providing feedback on how decisions do or do not reflect input received." (US DOE, 1995a)

Both the Fish and Wildlife Service and the Department of Energy examples illustrate a traditional view of public participation, one that seems to treat participation and involvement interchangeably.

The National Environmental Justice Advisory Council's (NEJAC) "Model Plan for Public Participation" (NEJAC, 2000) presents a more innovative and comprehensive view. As an advisory committee to the U.S. Environmental Protection Agency, the NEJAC's model includes "core values and guiding principles for the practice of public participation" (see Figure 6.2).

As these items indicate, the NEJAC believes that public participation should be multifaceted, meaningful activity that plays an important role in the environmental planning and decision-making process. The first seven statements in the NEJAC core values and guiding principles list come from the International Association for Public Participation (IAP2), as published in the spring 1996 issue of the IAP2 journal, *Interact* (NEJAC, 2000). The NEJAC emphasizes that any comprehensive public participation process addresses four critical elements: preparation, participants, logistics, and mechanics. These elements, along with the core values and guiding principles, suggest that public participation means much more than simple involvement.

Similarly, the IAP2 considers participation to be more encompassing than involvement. The IAP2 places involvement within its public participation spectrum. According to the spectrum, public participation extends from inform to consult to involve to collaborate to empower. Each represents an increased level of public impact (see Figure 6.3). The spectrum

Figure 6.2
Core Values and Guiding Principles for the Practice of Public Participation

1. People should have a say in decisions about actions which affect their lives.
2. Public participation includes the promise that the public's contribution will influence the decision.
3. The public participation process communicates the interests and meets the process needs of all participants.
4. The public participation process seeks out and facilitates the involvement of those potentially affected.
5. The public participation process involves participants in defining how they participate.
6. The public participation process communicates to participants how their input was, or was not, utilized.
7. The public participation process provides participants with the information they need to participate in a meaningful way.
8. Involve the public in decisions about actions which affect their lives.
9. Maintain honesty and integrity throughout the process.
10. Encourage early and active community participation.
11. Recognize community knowledge.
12. Use cross-cultural methods of communication.
13. Institutionalize meaningful public participation by acknowledging and formalizing the process.
14. Create mechanisms and measurements to ensure the effectiveness of public participation.

From: *The Model Plan for Public Participation*, National Environmental Justice Advisory Council. 2000. U.S. Environmental Protection Agency, EPA-300-K-00-001.

illustrates that the IAP2 views participation broadly, ranging from inform-and-educate activities to citizen decision making (see http://www.iap2.org for a complete version of the spectrum).

The IAP2 spectrum addresses aspects of both the process and outcome of decision making. As one moves across the spectrum, the process becomes more interactive, decision-making power is increasingly shared, and the decision is influenced or even dictated significantly by the public.

Figure 6.3
The IAP2 Public Participation Spectrum

Increasing Level of Public Impact
Inform → Consult → Involve → Collaborate → Empower

TRADITIONAL AND INNOVATIVE PUBLIC PARTICIPATION

As the previous discussion reveals, public participation is a broad term subject to varied approaches and interpretations. The variety of approaches can be framed as traditional and innovative, as Table 6.1 presents.

A Critical Element: Decision Space

Power sharing and participatory decision making are both indicative of *decision space*. The greater the decision space is, the greater the potential for meaningful public participation. Decision space is an important element that differentiates limited or traditional participation from more innovative and interactive participation.

Decision space is part of conflict assessment (Daniels & Walker, 2001). When assessing a controversial and complex environmental situation, issues of decision space and decision authority should be addressed. Assessment needs to reveal who has jurisdiction in the public policy decision situation and who has legal imperative to make or block a policy decision in that situation. Jurisdiction is related to decision authority—the individual or organization that has the legal or organizational duty to manage or regulate the situation.

Decision space stems from decision authority. Those parties with decision authority must clarify how much of the decision process and outcome they can share with other parties. The extent to which a decision authority can open up and share its decision-making process defines the decision space. For example, the Environmental Protection Agency, as a regulatory agency, enforces environmental laws such as the Clean Water Act and the Clean Air Act. The EPA can make decisions about clean water and clean air issues and impose those decisions on affected parties, or the EPA can invite affected and interested parties to work with the agency to determine how clean water and

Table 6.1
Comparing Traditional and Innovative Public Participation

Element	Traditional	Innovative
Goal	Information gathering and feedback	Fair, inclusive process; respectful interaction; mutual gains outcome
Decision space	Low, limited, vague	Significant and clear
Decision authority	Rigid	Flexible
Power	Centralized	Shared
Valued knowledge	Technical	Integrated; technical and traditional
Communication philosophy	Command and control	Dialogue and deliberation; Inquiry and advocacy
Communication activity	Inform and educate; gather feedback	Interaction, mutual learning, idea development and refinement
Access	Structured and controlled by the decision authority	Multifaceted, open, and inclusive; possibly designed by parties
Negotiation	None likely without appeals or litigation	Fostered; mutual gains interaction
Primary methods	Public hearings, comment letters, open houses, Web sites	workshops, roundtables, forums, dialogues,
Collaborative Potential	Low	Potentially high
Prospects for consensus	Not likely or sought	Possible
Measure of success	Quantitative; number of participant contacts	Qualitative; quality of participants' interaction and contributions
Democracy?	Government and elites	Governance and citizens

clean air standards can best be met. In the latter case, the EPA creates meaningful decision space while retaining its decision authority.

Sharing decision space involves sharing a form of power. Although the deciding agency retains its authority by law to make the decision (e.g., under NEPA a forest supervisor signs a record of decision), citizens can participate actively in the construction of that decision. Meaningful decision space is critical to a meaningful and innovative public participation process. Traditional public participation processes do not necessarily include any shared decision space. Any agency can consult with the public (e.g., invite comments in writing or at a hearing) without any assurance of how those comments might be part of the decision process. A traditional public participation process may embody a decision space facade.

COMMAND AND CONTROL VERSUS OPEN COMMUNICATION

Traditional public involvement approaches, Wondolleck and Yaffee explain, "usually provide highly controlled, one-way flows of information, guard decision-making power tightly, and constrain interaction between interested groups and decision makers" (2000, p. 104). Traditional methods for public participation are often part of a strategy of command and control (Weber, 1998), in which political decision making and technical information are guarded, and centralized power and hierarchy are maintained. Weber calls this strategy a conflict game that shuns pluralism and collaboration. While this view may seem quite strong, traditional public participation approaches are consistent with a planning strategy that is consultative rather than collaborative, one that emphasizes highly controlled communication rather than open interaction.

Consultation refers to those activities that involve parties in the environmental or natural-resource policy decision-making process without sharing any aspect of the decision itself. Whereas consultation is a legitimate and viable decision-making strategic option, it retains control of institutional power, decision making, and formal communication. It accords privilege to technical knowledge and does not accommodate traditional knowledge well (Brick & Weber, 2001). A consultative strategy is not collaborative, nor does it involve any form of consensus. Traditional public participation is consultative; its basic activities are information gathering and feedback. As Daniels and Walker explain:

> When a decision authority seeks input from other parties, it will invite feedback on its terms. The decision authority might present a range of possible alternative decisions or propose a specific action and then seek the reactions of other parties, such as those likely affected by the decision. The decision authority may ask for ideas as it begins a planning process. In either case, the decision authority

> provides opportunities for participation in the decision situation without participating in the process of decision making itself. (2001, p. 71)

Traditional public participation activities, such as issue scoping meetings, public hearings, letter-writing comment periods, open houses, and the like stem from and maintain decision-maker control. These techniques provide people and organizations with opportunities to communicate their concerns to a decision authority. The techniques seek to "inform and educate" and "invite feedback" while offering no guarantee of meaningful citizen input. In these settings, citizens do not know if and how their ideas will be used. Whether or not their comments influence the decision may depend on the benevolence of the decision authority (Daniels & Walker, 2001).

Consultation approaches—traditional public participation activities—have attributes that appeal to administrators and decision makers who want to limit decision space while maintaining power and control. Consultation methods generally provide citizens with relatively easy access to the policy process; almost anyone can write a comment letter. Open houses and public hearings are generally well publicized, and the behavioral expectations and opportunities are predictable. Consultative methods may be efficient in terms of time and expense. The Forest Service's fast track public participation approach to the roadless forest lands question seems quite efficient as measured by time.

The consultative strategy and its traditional public participation techniques give rise to the various limitations: the uncertainty over how citizen comments are used, the limited impact that comments have on the outcome, the quasi-arbitration authority of the deciding official (the agency as arbitrator), the formality of the communication environment, and the correspondent perceptions of a zero-sum game (Walker & Daniels, 2001).

Communication activity in consultative processes reflects a philosophy of "command and control," limiting the quality of the information that the agency receives. For example, to comment at a hearing-type meeting, a participant must speak for the record, which is often equivalent to making a short speech into a microphone before a relatively large assembly. Given the proportion of people in whom such public speaking produces anxiety, it is likely that the quality and quantity of the comments is reduced by such a formal protocol, and that only the most motivated people will overcome their fears and address the group. As a result, the comments tend to be more extreme than they might be in a setting where dialogue is more natural.

In contrast, innovative public participation features open constructive communication. One significant form of innovation is collaboration. Wondolleck and Yaffee observe that

> [T]he most successful collaborative efforts fostered two-way, interactive flows of information, and decision making occurred through an open, interactive process rather than behind closed agency doors. Such efforts actively involved people throughout a planning or problem-solving process so that they learned together, understood constraints, and developed creative ideas, trust, and relationships. Direct face-to-face interaction between stakeholders and decision-making authorities was critical. (2000, p. 105)

There are many collaborative methods and approaches such as participatory decision making, search conferencing, and collaborative learning (see Daniels & Walker, 1999, for a review). These and other collaboration methods have a variety of common attributes such as multiple stages; constructive, open, civil communication, generally as dialogue; a focus on the future; an emphasis on learning; and some degree of power sharing and leveling of the playing field (Daniels & Walker, 1999).

Environmental planning and decision making processes that employ innovative public participation will foster both dialogue and deliberation. Dialogue encourages learning, but deliberative interaction is necessary for making decisions. As parties listen, learn, and deliberate, they will engage in both inquiry and advocacy (Daniels & Walker, 2001; Senge, 1990). Such open, constructive communication will enhance the potential for collaboration and innovation.

Collaborative Potential

Any party, whether the decision authority, a key stakeholder, or a citizens' group, that seeks to implement an innovative public participation strategy will likely perceive collaborative potential. Collaborative potential can be defined as the opportunity for parties to work together assertively to make meaningful progress in the management of controversial and conflict-laden policy situations. This perception is based on two factors. First, the party believes that there is a possibility for meaningful, respectful communication interaction between the disputants. Second, the party believes that a mutual gain or integrative outcome is possible; that is, that the fundamental structure of the conflict or decision situation offers the potential for both or all sides to achieve more of their objectives than would be likely in some other venue (Lewicki, Saunders, & Minton, 1999).

As a public participation strategy, collaboration differs considerably from the traditional model of open houses, public hearings, and comment periods. Some key aspects of collaboration that clarify these differences are: (1) it is less competitive, (2) it features mutual learning and fact-finding, (3) it allows underlying value differences to be explored, (4) it resembles principled negotiation, focusing on interests rather than positions, (5) it allocates the responsibility for implementation across many parties, (6) its conclusions are generated by participants through an interactive, iterative, and reflexive process, (7) it is often an ongoing process, and (8) it has the potential to build individual and community capacity in such areas as conflict management, leadership, decision making, and communication (Daniels & Walker, 2001).

Collaborative potential can be assessed via any of a number of frameworks, including the conflict map (Wehr, 1979); the conflict dynamics continuum (Carpenter & Kennedy, 1986/1988); the Progress Triangle (Daniels & Walker, 2001); or any other suitable approach. Regardless of the framework employed, the assessment should help the analyst determine: (1) the current potential for collaboration, and (2) the extent to which certain aspects of the situation need to be changed to establish good potential for collaboration. There is no formula to this assessment process. Rather, the analyst has to assess the situation as comprehensively as possible given available resources to do so, such as time, access to people for interviews, review of documents, and so on. In environmental policy conflict situations, though, the willingness of parties to try to work together and the degree of decision space the relevant decision makers are willing to share are key factors.

THE ROADLESS INITIATIVE PUBLIC PARTICIPATION STRATEGY: TRADITION ENSCONCED

The public participation component of the Roadless Initiative featured hundreds of public meetings, a well-developed Web site, and hundreds of thousands of comment letters. Former Forest Service Chief Mike Dombeck has cited all this as evidence of an open process. It stands instead as evidence of a traditional process, perhaps even serving as a poster example of traditional, conventional, command-and-control public participation.

DECISION SPACE

The public first learned about the Roadless Initiative via a statement that communicated something profound—no meaningful decision space. NEPA calls for an environmental impact statement to include a "no action" alternative, yet the message in President Clinton's October 1999 address was clear—the only viable alternative was action.

In his speech at the George Washington National Forest, after invoking the memory of Theodore Roosevelt's public lands vision, President Clinton told the gathered audience:

> Today, we launch one of the largest land preservation efforts in America's history to protect these priceless, backcountry lands. The Forest Service will prepare a detailed analysis of how best to preserve our forests' large roadless areas, and then present a formal proposal to do just that. The Forest Service will also determine whether similar protection is warranted for smaller roadless areas that have not yet been surveyed.
>
> Through this action, we will protect more than 40 million acres, 20 percent of the total forest land in America in the national forests—from activities such as new road construction, which would degrade the land. We will ensure that our grandchildren will be able to hike up to this peak, that others like it across the country will also offer the same opportunities. We will assure that when they get to the top they'll be able to look out on valleys like this, just as beautiful then as they are now. (Clinton, 1999b)

In his memorandum to the Secretary of Agriculture released on the same day, President Clinton wrote that

> it is time now . . . to address our next challenge—the fate of those lands within the National Forest System that remain largely untouched by human intervention. . . . Accordingly, I have determined that it is in the best interest of our Nation, and of future generations, to provide strong and lasting protection for these forests, and I am directing you to initiate administrative proceedings to that end.

Clinton then proposed specific action. "I direct the Forest Service to develop, and propose for public comment, regulations to provide appropriate long-term protection for most or all of these currently inventoried 'roadless' areas, and to determine whether such protection is warranted for any smaller 'roadless' areas not yet inventoried" (Clinton, 1999a).

These statements suggested that the decision to protect millions of National Forest acres as roadless had already been made; what remained was working out the details and meeting the requirements of relevant planning statutes such as NEPA and the National Forest Management Act. In his speech and memorandum, President Clinton, as the nation's chief executive, proposed that over 40 million acres of national forest land be designated as roadless and protected. He directed the Forest Service, a Department of Agriculture agency, to develop a plan to do so. After hundreds of scoping meetings held in November and December 1999, the Forest Service released its draft environmental impact statement in May, 2000. The DEIS specified four alternatives: a no-action alternative (as required by law) and three alternatives that protected unroaded areas and restricted or prohibited timber harvest.

Table 6.2 compares President Clinton's statements with the DEIS preferred alternative, the Final Environmental Impact Statement (FEIS) preferred alternative, and the Record of Decision (ROD). As the information in Table 6.2 indicates, changes occurred during the planning process. President Clinton did not refer specifically to the Tongass National Forest in his October 13, 1999, communication. The DEIS proposed postponing action on the Tongass National Forest until 2004. The ROD included the Tongass in the new roadless policy.

Although there are modest changes across the DEIS, FEIS, and ROD on the issues of road construction and timber harvest prohibition, all are very consistent with President Clinton's initial message. As has been noted earlier in this chapter, Clinton tells the Secretary of Agriculture that he has determined that "strong and lasting protection for these [National] forests" is essential and that a planning process to accomplish this goal should begin immediately. The President articulated the baseline for the preferred alternative.

Command and Control?

The limited decision space and the short time line combined to discourage any significant innovation in the Roadless Initiative public participation strategy. Still, the Forest Service did conduct an ambitious public involvement campaign. In July 2000 testimony before Congress, Forest Service Deputy Chief Jim Furnish reported that:

Table 6.2
From President Clinton to the Final Rule

	President Clinton's statements (October 13, 1999)	Proposed Rule & DEIS Preferred Alternative (May 9, 2000)	FEIS Preferred Alternative (November 13, 2000)	Final Rule & Record of Decision (January 5, 2001)
Road Construction Prohibitions	We will protect more than 40 million acres of national forest land from activities, such as new road construction, which would degrade the land. (speech) I direct the Forest Service to develop, and propose for public comment, regulations to provide appropriate long-term protection for most or all of these currently inventoried "roadless" areas, and to determine whether such protection is warranted for any smaller "roadless" areas not yet inventoried. (memo)	Prohibits new road construction or reconstruction in the unroaded portions of inventoried roadless areas on National Forest System lands, except: • To protect health and safety threatened by a catastrophic event; • To conduct environmental clean up; • For valid existing rights; • To prevent irreparable resource damage by an existing road.	Prohibits new road construction and reconstruction within inventoried roadless areas on National Forest System lands, except: • To protect health and safety threatened by a catastrophic event; • To conduct environmental clean up; • To allow for reserved or outstanding rights provided for by statute or treaty; • To prevent irreparable resource damage by an existing road; • To rectify existing hazardous road conditions; • When a road is part of a Federal Aid Highway project.	Same as the Preferred Alternative, plus • Road construction may be allowed in conjunction with the continuation, extension, or renewal of a mineral lease on lands that are under lease or for new leases issued immediately upon expiration of an existing lease.

(*continued*)

Table 6.2
From President Clinton to the Final Rule (*continued*)

	President Clinton's statements (October 13, 1999)	Proposed Rule & DEIS Preferred Alternative (May 9, 2000)	FEIS Preferred Alternative (November 13, 2000)	Final Rule & Record of Decision (January 5, 2001)
Timber Harvest Prohibitions	This initiative should have almost no effect on timber supply. Only five percent of our country's timber comes from national forests. Less than five percent of the national forests' timber is now being cut in roadless areas. We can easily adjust our federal timber program to replace what we might destroy if we don't protect these 40 million acres. (speech)	Timber harvest is allowed, as long as such activity does not include constructing any new roads or reconstructing any existing roads [allows helicopter logging].	Prohibits timber harvest, except for purposes to maintain or improve roadless characteristics and: • To improve threatened, endangered, proposed or sensitive species habitat; • To reduce the risk of uncharacteristic wildfire effects; • To restore ecological structure, function, processes, or composition.	Prohibits cutting, sale, and removal of timber in inventoried roadless areas, except: • For the removal of small diameter trees that maintains or improves roadless characteristics; and • To improve threatened, endangered, or sensitive species habitat; • To maintain or restore ecosystem composition and structure, such as reducing the risk of uncharacteristic wildfire effects.
Tongass National Forest	Not addressed	Decide to include/ not include the Tongass NF in 2004	The above prohibitions go into effect in 2004	Prohibitions apply immediately to the Tongass NF

Note. Adapted from *Memorandum for the Secretary of Agriculture*, by W. J. Clinton, 1996b, USDA Forest Service Roadless Conservation Web site. Available: http://www.roadless.fs.fed.us/documents.

> The Forest Service is conducting an unprecedented public outreach effort to solicit public comments on the roadless area conservation proposal. . . . During [the] scoping period the Forest Service held more than 185 public meetings across the country and received over 500,000 comments.
>
> On May 9, 2000, the agency started the just concluded public comment period on the proposed rule and draft environmental impact statement. The agency held an additional 424 meetings, including at least two meetings on every national forest across the country—one to explain the proposal and one to hear public comments. In total, the Forest Service has held more than 600 meetings on this proposal. The agency has posted its draft environmental impact statement and maps on the Internet and has distributed over 43,000 copies of the DEIS, almost 6,000 CD versions of the DEIS, and over 50,000 copies of the DEIS summary to interested individuals, local governments, and other agencies. In addition, the agency also distributed copies of the DEIS to over 10,000 public libraries across the country. (Furnish, 2000)

The Forest Service has measured the success of its public participation strategy quantitatively, primarily by the number of meetings it held, the number of people who attended those meetings, and the number of comment letters the agency received. The Forest Service's roadless policy Web site includes a table that lists, by region, the number of meetings held during the DEIS review period, the number of attendees, and the percentage who talked formally. This table is accompanied by a list of public involvement highlights, including:

- Information meetings held in late May and early June were designed to give members of the public a chance to review the Proposed Rule and Draft Environmental Impact Statement with local Forest Service representatives. Public comment meetings in late June and early July allowed citizens who wished to comment verbally on the proposal to do so for the public record.
- More than 440 meetings nationwide—more than 230 info meetings, more than 200 comment meetings. This included Washington DC–area meetings, regional meetings, forest meetings and some district-level meetings.
- Total for information and comment meeting attendance is estimated near 25,000 based on registration information from forests.

- Wide range in attendees at public info meetings—from a handful (2–3) to 100+. Several forests had zero attendees. Comment meeting participants ranged from 1 to 370.
- Comments have been received by letter, postcard, e-mail, telefax and through transcripts recorded at public comment meetings. More than 1 million individual responses have been received—more than 95 percent were simple postcards or other form letters—the rest were individual letters. (Roadless Area Conservation, 2000)

By employing a traditional public participation strategy, by pursuing a short planning time line, and by measuring public participation success quantitatively, the Forest Service became "agency as arbitrator." As Walker and Daniels explain:

> The game theoretic incentives embedded in traditional public participation increase the likelihood of extreme behaviors. Agency decision making resembles conventional arbitration with the deciding official acting as an arbitrator, and different public groups making their cases before the arbitrator in an effort to affect the decision. Theoretical research into the incentives created in conventional arbitration formats shows clearly that the incentives for the participants is to state extreme demands and use volatile rhetoric, because the assumption is that the decision maker/arbitrator will somehow split the difference between the different groups. In order to move the decision in their desired way, each group must be more forceful and compelling than the others. The old maxim of "the squeaky wheel getting the grease" is applicable to these incentives; the group that squeaks the loudest gets the influence on the outcome. Such "squeaks" may have little to do with scientific and technical knowledge applicable to the situation; the agency as arbitrator may respond to "pressure politics" in ways that diminish scientific as well as citizen interests. (2001, p. 258)

I witnessed lobbying, posturing, extreme behavior, and volatile rhetoric at the various Roadless Initiative meetings I attended. At one information meeting, the local off-road vehicle community attended in great numbers, sat together, and insisted that Forest Service officials address their specific concerns. The DEIS public comment meetings were conducted like public hearings. At the Eugene, Oregon meeting, speakers representing strong environmental values outnumbered speakers voicing support for off-road vehicle use or timber harvest values by 10 to 1. A number of the environmental values speakers (all of whom advocated more roadless area designations including the Tongass National Forest) were young women holding infants. They spoke passionately on behalf of future generations.

Organizations nationwide encouraged their members to attend meetings, and more important, to send in comments. Numerous organizational Web sites included form letters and addresses. As noted earlier, the Forest Service has reported on its Roadless Initiative Web site that 95 percent of the comment letters received were either postcards or form letters. Many organizations, from environmental coalitions to industry and off-highway vehicle (OHV) groups, posted addresses and sample letters on Web sites and in newsletters. Organizations also mobilized their members to turn out in force at the many public meetings. The status of Tongass National Forest likely changed in part because environmental groups mobilized their members to write and attend meetings better than other groups.

COLLABORATIVE POTENTIAL?

It is no surprise that there was no meaningful collaborative potential in the Roadless Initiative planning process. President Clinton established the basic outline and foundation of the preferred alternative and final action in his October 1999 memorandum and speech. There was little for citizens to influence beyond adding land (particularly the Tongass National Forest) and nothing to negotiate. The structure of the planning process and public participation activities negated any possibility of mutual-gain outcomes and integrative solutions. The hierarchical, centralized, and time-bound approach on the roadless issue rendered any collaborative strategy unlikely.

Wondolleck and Yaffee (2000) remark that "one simple message from many of the successful collaborative initiatives we examined is that involving the public early and often throughout a decision-making process is more likely to result in more effective decisions and produce satisfied stakeholders" (p. 103). They contrast collaboration with traditional public participation mechanisms through which "agency officials try to convince the public to accept their plans" and interest groups "grandstand" (p. 104).

AN ALTERNATIVE MODEL: THE SEVENTH AMERICAN FOREST CONGRESS ROUNDTABLES

This discussion of the Roadless Initiative public participation strategy suggests two important questions. Were there any viable alternative public participation strategies? Could a national policy be addressed collaboratively? Community-based collaborative groups have emerged during the past few years to address national resource and environmental controversies. Although some of these groups have been quite productive, critics have contended that local groups should not take action inconsistent with national interests (see Wondelleck & Yaffee, 2000; and Daniels & Walker, 2001 for discussions). Some critics of the Roadless Initiative have argued the opposite: that a national policy needs to account for local interests and provide opportunities for local participation (Lance, 2001; Pfleger, 2001).

Could the Roadless Initiative have been addressed in a way that accounted for both national and local interests? The Seventh American Forest Congress, held in 1995 and 1996, may offer an answer. For a week during February 1996, over 1,400 citizens met in Washington DC to discuss the future of the United States' forest lands. As participants in the Seventh American Forest Congress, these citizens worked through the week as members of ten-person roundtables. Reflecting a diversity of values and variety of interests, they discussed all issues of forest ecosystems: the aesthetic as well as the biological, the cultural as well as the physical, the economic as well as the spiritual (Cleland, 1996).

From the 147 roundtables emerged vision elements and principles to which congress participants responded with votes of "green" (agreement), "yellow" (unsure/need more information), and "red" (disagree). For example, a vision element specifying that "in the future our forests will be sustainable; support biological diversity; maintain ecological and evolutionary processes; and be highly productive" received 65 percent "green" votes and only 9 percent "red" votes. In contrast, the vision statement "in the future our forests will have decreased demand placed upon them as consumption of forest resources is aligned with sustainable forest management" generated 17 percent green votes and 58 percent red votes (American Forest Congress, 1996; Forest Congress Information Center, 1998).

What culminated in a national congress on forestry issues began months earlier as local roundtables and collaborative meetings. Between June 1995 and February 1996, 51 roundtables took place in 35 states. Each roundtable employed trained facilitators and lasted at least one day. Roundtable participation was open to anyone interested in forest policy issues. Thirty-nine collaborative meetings were conducted by interested parties (e.g., communities of interest, industry groups, citizen organizations). Both the roundtables and collaborative meetings were asked to discover common ground, but not necessarily consensus, on a range of forest policy issues (Forest Congress Information Center, 1998; OD Corp, 1996).

The Seventh American Forest Congress was held out of both frustration and failure with the politics of natural resources management (Little, 1996). At both a Yale Forestry Forum and Arbor Day meeting in Nebraska, academic group, environmental group, and timber industry leaders emerged as the core of the Congress Board of Directors. They created a core vision that called for significant involvement "of very diverse groups talking with each other in a process which encouraged brand new forms of much more open communication and ways of working together" (OD Corp, 1996, p. 1).

By most accounts, the Seventh American Forestry Congress made progress on forest policy issues and controversies (e.g., Little, 1996; OD Corp, 1996). The local and national gatherings also, in a modest way, improved relationships among varied stakeholders. The congress's success was grounded in part in a number of important process design features. First, many local roundtables and all of the national meeting tables involved people with diverse backgrounds, values, and worldviews. Second, at the national meeting, group process guidelines accounted for the inevitability of conflict and the importance of direct, constructive, and respectful interaction. Third, all meetings were guided by trained facilitators. Last, congress leaders and facilitators did not take substantive positions on the issues; congress participants were empowered to work through the issues as they deemed appropriate (American Forest Congress, 1996; OD Corp, 1996). The Congress seemed to make progress on the challenge of incorporating and integrating local and national interests (Little, 1996).

Could a similar design have been a component of the Roadless Initiative public participation strategy? Either prior to the Notice of Intent or during as part of the scoping process, national forests, national grasslands, and Bureau of Land Management districts could have convened local roundtables on all aspects of the Roadless Initiative along the lines of the Forest Congress, the democratic forum model (Shannon quoted in Burns, 2001), collaborative learning (Daniels & Walker, 2001), or some other similar approach. These roundtables, directed by trained facilitators, could have emphasized mutual learning and dialogue as a part of collaborative interaction. Following the local meetings, a national symposium could have been held, with participants from the local roundtables as well as any other interested parties encouraged to participate.

CONCLUSION

The title of this chapter poses the question: Did the Forest Service's public-involvement approach on the Roadless Initiative render "public participation" an oxymoron? The Forest Service did hold hundreds of public meetings, performed content analysis on more than one and one-half million comment letters and cards, and developed a very comprehensive and polished Web site. In fairness to the Forest Service staff members who were asked to carry out the Clinton-Glickman-Dombeck directive, the public participation activities were not oxymoronic. They were, though, prototypes for traditional public participation. Contrary to former Forest Service Chief Dombeck's claim, this strategy was not "open." Rather, the strategy illustrated ironies and an opportunity lost.

Two ironies stand out. First, at a time when the Forest Service has been emphasizing collaboration, the Roadless Initiative public participation strategy was quite the contrary. During the Roadless Initiative planning process, the Forest Service released the report of its three-year task force on collaborative stewardship. During this period, a special committee of scientists, appointed by the Secretary of Agriculture, developed new Forest Service planning regulations that emphasize collaboration. Despite the apparent importance the Forest Service has accorded collaboration, apparently the Roadless Initiative did not warrant it.

A second irony is that the Forest Service Roadless Initiative planners do not seem to recognize this collaboration contradiction. In its January 2001 "Roadless [PowerPoint] Presentation," the Forest Service recognizes the significance of collaboration in the new planning regulations, claims that the planning process uses science and local involvement, and asserts that public interest in the planning process was "unprecedented" (Roadless Area Conservation, 2001a, p. 2). This view certainly differs from that of an Idaho industry and motorized recreation coalition that has sued the federal government to stop the implementation of the roadless policy. A spokesperson for the coalition has stated that "the Forest Service bypassed the forest planning process, ignored forest plans and intends a top-down mandate" (Associated Press, 2001).

The opportunity lost may seem obvious. Here was a chance for the Clinton administration and the Forest Service to model an innovative process for involving citizens and interest groups in a national policy. By not doing so, planners contradicted the Forest Service emphasis on collaboration, reasserted an emphasis on traditional public participation, and may have undercut an arguably good policy idea.

In a recent *American Forests* essay, Gray and Kusel observe that "community-based practitioners feel—after two decades of 'public involvement' in which their comments have been synthesized, coded, counted, considered too late, or taken out of context—they have had little or no impact around them" (1998, p. 28). Gray and Kusel add that citizens do not seek to "call the shots," but desire a more open process that includes local knowledge, a variety of perspectives, and a shared commitment to stewardship, reinvestment, and monitoring (pp. 28–29). Twarkins, Fisher, and Robertson (2001) note that Forest Service planners and citizens alike value the following in a forest planning process: transparency, partnership, stakeholder analysis and involvement, multiple methods, dialogue, and education. In a similar vein, Moote et al. (2001) assert that credible, community-based, collaborative process features inclusiveness, diversity, accessibility, transparency, mutual learning, collective vision, and adaptability. Although the Roadless Initiative public participation strategy included some of these characteristics (e.g., access and multiple methods), it exemplified a business-as-usual approach rather than innovation and civic deliberation.

The Roadless Initiative and subsequent Roadless Area Conservation Rule may be good policy. Certainly there is strong scientific evidence to support this claim. As of this writing the new Rule is on hold, subject to revisions under the Bush administration. The Forest Service is currently seeking more public comment, with a Lettermanesque list of "ten questions." Of these, question two asks about "Working Together. What is the best way for the Forest Service to work with the variety of States, tribes, local communities, other organizations, and individuals in a collaborative manner to ensure that concerns about roadless values are heard and addressed through a fair and open process?" (Roadless Area Conservation, 2001b). This chapter has provided one answer: not the way the Forest Service has conducted Roadless Initiative public participation.

CHAPTER 7

PUBLIC PARTICIPATION AND (FAILED) LEGITIMATION: THE CASE OF FOREST SERVICE RHETORICS IN THE BOUNDARY WATERS CANOE AREA

STEVE SCHWARZE

Issues of public participation come into sharp relief in rhetorics surrounding management of United States Forest Service lands. Although management decisions on the national forests generate public controversy for a variety of reasons, public criticism and defense of the agency's decisions consistently highlight the role of public participation in the decision-making process. To the extent that the agency implements decisions that run counter to the values and beliefs of citizens and interest groups, and appears to ignore appeals made by public advocates, questions about the value of public participation get raised in arenas of public discourse.

The persistence of these questions potentially erodes the legitimacy of the Forest Service in the eyes of citizens and interest groups. Criticism of the Forest Service's public participation efforts goes right to the heart of the question of legitimacy because the agency is indirectly a means or tool of the people. In a democratic society, public agencies garner legitimacy to the extent that their actions can be persuasively explained as expressing the will of the people. Hence, the repeated criticisms of the Forest Service's inability to structure adequate public participation strikes at the core of the service's legitimacy as a public agency.

To deflect these criticisms, the Forest Service's management plans attempt to produce legitimacy by highlighting the role that public participation plays in their decision-making process. But do these plans successfully produce legitimacy? Is it even possible for the plans to do so? In this essay, I address those questions and contend that the Forest Service's management plans are a rhetorical failure. The management plans function rhetorically to erode the legitimacy of the agency and its decisions. I support this contention by engaging in a case study of Forest Service plans developed to guide management of the Boundary Waters Canoe Area Wilderness (BWCAW) in northeastern Minnesota. The Boundary Waters area has been a site of contention since its designation as a roadless area in 1926, and the controversies over its management have consistently pitted Forest Service personnel against local citizens and legislators.

Although various explanations exist for why the Forest Service often finds itself in situations of controversy, a rhetorical perspective best explains how these controversies erode the legitimacy of the agency. Other accounts suggest that the Forest Service faces controversy because it is handcuffed by a multiple-use mandate, because it implements decisions with significant material impacts on vocal interest groups, and because citizens have gained more opportunities for participation in the past century. But the rhetorics by which the service justifies its decisions regarding these controversies are not equipped to bolster the service's legitimacy. Forest Service management plans, much like other planning documents developed by federal land-management agencies, are documents that fail rhetorically. The plans fail because they must bear the burden of both description of the process and legitimation of the process. Whereas the plans can successfully describe how the agency fulfills the demands of bureaucratic rationality involved in developing those plans, it is less suited to legitimize decisions in the face of conflicting citizens and interest groups. The plans offer an institutional rationale rather than a persuasive public rhetoric.

The argument of this chapter proceeds in the following steps. First, I provide the theoretical context for the chapter by discussing important scholarship in environmental communication that address the representation of public participation in agency rhetorics. I suggest that my case study both confirms and problematizes the conclusions drawn by Killingsworth and Palmer in their study of environmental impact statements (1992). Second, I examine the Forest Service management plans for the Boundary Waters, showing how the plans represent public participation in ways that invite criticism from citizens and interest groups. Third, I support my claim that the plans erode legitimacy by outlining some criticisms of the most recent management plan. After establishing how the plans erode legitimacy, I offer an explanation of why that happens, showing how the rhetorical constraints on the plans inhibit the consideration of legitimacy as a potential rhetorical effect. This leads into a conclusion that speculates on rhetorical alternatives that could both strengthen public participation and restore legitimacy to the Forest Service.

RHETORICAL REPRESENTATIONS OF PUBLIC PARTICIPATION

The representation of public participation in agency rhetorics dominates two important essays in rhetorical studies. First, Peterson and Horton's (1995) essay on the conflict over conservation of golden-cheeked warbler habitat in Texas shows the absence of a significant group's voice in public discourses. In their analysis of the conflict, Peterson and Horton claim that the voices of environmental activists and agency personnel dominate public discourses, while the voices of ranchers involved in habitat protection are barely heard. This absence exacerbates mistrust of the agency and diminishes the possibility of ranchers and agency personnel collaborating to produce mutual beneficial solutions to the habitat conflict: "By restricting opportunities to participate in the public sphere, the United States Fish and Wildlife Service (USFWS) has assumed an adversarial position relative to an audience whose cooperation they desire" (p. 140). In Peterson and Horton's view, the USFWS's failure to actively incorporate the knowledge and concerns of local ranchers in the conservation process explains the ongoing conflict between ranchers and the agency.

Peterson and Horton's essay identifies the most glaring problem related to public participation: the absence or explicit exclusion of certain voices. Peterson and Horton not only retrieve the ranchers' voice in their analysis, but also go on to demonstrate how the mythemes of the ranchers contain potential points of affiliation with agency and environmental discourses. Their essay provides a useful example of how rhetorical analysis can aid in the resolution of environmental conflict between agencies and their constituents.

However, the mere representation of multiple voices within agency decision making does not eliminate the barriers to meaningful public participation. Indeed, often a more complex problem emerges: the marginalization of certain voices within a broader rhetoric of participation. Despite proclamations of significant public participation, agency rhetorics often function to control and exclude robust participation. Killingsworth and Palmer, whose essay (1992) on environmental impact statements shows how public participation gets thwarted in the production of those documents, explain this problem in terms of instrumental rationality.

Employing Habermas's distinction between instrumental and communicative rationality, Killingsworth and Palmer argue that a "rhetoric of instrumentalism dominates the major documents produced by administrative government" (p. 163), and that this instrumentalism shows how little influence public participation seems to have on agency decision making. The relationship between instrumental rationality and public participation manifests itself in the Environmental Impact Statement (EIS) in two ways. First, the objective style of the documents preserves the status of agency experts and excludes voices that speak about the environment in ways that deviate from the agency's way of framing the issues. Second, where the EIS does include alternate voices, it merely reprints public comments without showing any substantive connections between those voices and agency decisions, blunting the impact of those voices. Hence, the voices get marginalized within a rhetoric of participation: "Though they may draw upon the conventions of a democratic discourse that is open to information from diverse sources, the aim of instrumental documents is never to treat deviant discourses with respect but always merely to take note of them, to record them, and ultimately to treat them as 'noise' in the system, which needs to be ignored or expunged" (p. 166).

Killingsworth and Palmer's analysis of EIS produced by the Bureau of Land Management (BLM) in New Mexico goes on to demonstrate how "deviant" rhetorics were marginalized and essentially ignored by the agency. Although the final EIS included citizens' written responses to the draft EIS, those responses were published in their original state, their "native roughness," in contrast to "the technical quality of the BLM-produced portion of the document" (p. 186). Moreover, the responses themselves indicated a "tacit recognition that their voices will have little effect—rhetorical or real" (p. 186). Killingsworth and Palmer conclude that although the incorporation of these responses offers appearance of a democratic discourse, the content and placement of the responses undermines the democratic impulse. Ultimately, the EIS functions primarily to rationalize a plan of instrumental action and fails to generate communicative action among diverse groups (p. 190).

A similar process of marginalization appears to have occurred in the management of the Boundary Waters Canoe Area Wilderness in Minnesota.

Like the cases in Texas and New Mexico, some citizens and interest groups claim that a federal agency (in this case, the Forest Service) has ignored local voices in the decision-making process on the Wilderness. As in other instances, perception of the agency's lack of interest in public participation has heightened the adversarial character of the relationship between some local residents and the agency. Thus, this chapter shares the concerns raised by Peterson and Horton, and Killingsworth and Palmer.

By examining these concerns in the case of the Boundary Waters, this chapter both confirms and problematizes the critique of agency rhetoric offered by Killingsworth and Palmer. On one hand, this chapter confirms that agency rhetorics can be characterized in terms of an objective style. In this case, however, that style is most apparent in the way that agency documents describe public participation. The Forest Service's description of public participation objectifies the participation process, focusing on the amount of comments gathered and the number of hearings held. This quantification of the public participation process, I suggest, helps make the process transparent, both in the sense of making the process clear to the reader and emptying it of substantive content. Agency rhetoric casts public participation as a set of basically anonymous objects to be manipulated for instrumental purposes, rather than as means by which multiple subjects can advocate divergent, substantive positions.

On the other hand, this chapter intends to problematize Killingsworth and Palmer's argument that agency rhetoric ultimately fails insofar as it serves a narrow, covert purpose: defending agency action. I agree that the EIS and other planning documents serve this purpose, but I disagree that this purpose itself marks a failure of the discourse. Instead, I contend that the documents fail to adapt to the general public audience, and in the process the agency fails to move beyond justification within an institutional context to produce legitimacy among the broader set of citizens and interest groups with a stake in the controversy. To contextualize this disagreement, I return to Killingsworth and Palmer's argument about the covert purpose of the EIS.

Their final critique is that the EIS originally contained the possibility of serving as a vehicle of communicative action, but agency motives turned the EIS into a tool of defense. They refer to several studies that argue the EIS has been co-opted for the instrumental purpose of defending agency actions from legal attack. Even though laws and regulations require public participation in the process of constructing the EIS, the production of that document has not generated the sort of participation necessary for realizing communicative rationality. Instead, the document becomes a tool with which agency can show that they have fulfilled all their legal obligations. As Killingsworth and Palmer put it, "The EIS becomes a means of proof, or certification, or demonstration. The EIS is not intended to inform action but to forestall action—legal action against the agency in question" (p. 189).

Although I agree that the EIS serves this function, I believe that a critique of the EIS on this ground is unfair. It is difficult to see how a document required by and produced within a complicated matrix of legal obligations could avoid giving significant attention to steps the agency has taken to fulfill those obligations. In my view, the problem is not that the document's covert purpose is to defend the agency from legal action. Rather, the problem is that the document fails to give sufficient consideration to citizens and interest groups as potential audiences. In doing so, the texts foreclose other rhetorical possibilities for generating legitimacy.

These other possibilities are clarified by W. Richard Scott's institutional perspective on legitimacy (1995). Scott argues that theorists have identified three broad dimensions or "pillars" that support institutions—regulative, normative and cognitive—and that each pillar implies a different basis for legitimacy. From a rhetorical perspective, talking in terms of a basis for legitimacy suggests that legitimacy may be characterized in terms of an argument; that is, a claim in need of support in relation to some audience. Thus, the different bases for legitimacy identified by Scott imply different forms of appeal that might be used to generate legitimacy. In Scott's words, "From an institutional perspective, legitimacy is not a commodity to be possessed or exchanged but a condition reflecting cultural alignment, normative support, or consonance with relevant rules or laws" (p. 45). In other words, legitimacy is not an inherent property of an institution but a rhetorical effect produced through three broad types of appeal: cognitive (appeals to cultural categories of common sense), normative (appeals to values and norms), and regulative (appeals to codified rules and laws).

The differences between these bases for legitimacy helps clarify and explain the failure of the Forest Service rhetorics in producing legitimacy. As I claimed in the introduction of this chapter, Boundary Waters management plans offer an institutional rationale rather than a persuasive public rhetoric. In Scott's terms, this means that the Forest Service's plans are constituted by a thoroughly regulative view of institutions and legitimacy, and that they fail to capitalize on normative and cognitive modes of appeal that might generate legitimacy. As Scott claims, the regulative basis for legitimacy emphasizes conformity to rules and laws: "Legitimate organizations are those established by and operating in accordance with relevant legal or quasi-legal requirements" (p. 47). To the extent that the plans emphasize how the Forest Service is operating in accordance with the laws and regulations surrounding management, they fall squarely within this regulative view articulated by Scott. But this regulative view may not be held by other relevant audiences; instead, the values that undergird a decision and the cultural categories employed in the plans may not resonate with citizens and interest groups. To the extent that the Forest Service plans remain embedded within a regulative view of institutions, they fail to realize the potential of other means of generating legitimacy.

Indeed, the case of the Boundary Waters shows that agency documents do not even do a good job of achieving their covert purpose—defending agency action—precisely because the agents of legal action are citizens and interest groups. As I will show, regulative appeals do little to generate legitimacy in the eyes of these groups. If the agency cannot produce discourse that generates legitimacy among interested publics, then they will continue to face administrative appeals and lawsuits from those publics. Now I will turn to an analysis of the Boundary Waters management plans to show the rhetorical conventions by which the agency supports its decisions.

BOUNDARY WATERS MANAGEMENT PLANS AND THE RHETORIC OF PARTICIPATION

This section of the chapter performs a textual analysis of the Boundary Waters management plans. I use the term *management plans* to refer to the collective set of documents produced at the end of the planning process, documents that include environmental impact statements of the kind described by Killingsworth and Palmer, Records of Decision (ROD), and Implementation Schedules. This analysis demonstrates how those documents turn a blind eye to citizens and interest groups. They do so in two significant ways: by making public participation transparent, and by emphasizing accountability to legislative and judicial mandates. In contrast to Killingsworth and Palmer, I do not criticize these rhetorical conventions merely because they defend the agency's actions. Rather, I suggest that the texts' representations of public participation and emphasis on justification within a regulative framework (i.e. decision making constrained by laws and rules) deflects attention from the material and normative concerns of citizens and interest groups. In this way, the rhetorical function of legitimating the agency gets ignored. Although the document could function to promote the agency as responsive to public concerns, its treatment of public participation and the focus on legal issues produces the impression of an institution that has little concern for public claims.

THE 1981 PLAN TO IMPLEMENT THE BOUNDARY WATERS ACT

The 1981 Implementation Plan (USDA, 1981) goes to great lengths to make the public participation process transparent. The final section of the introduction makes brief mention of public involvement in identifying needs and proposing actions since the passage of the Boundary Waters (BW) Act, reinforcing the Forest Service's responsiveness to the people. However, it is in the appendixes to the plan where the most direct references to public participation are made. Appendix B provides a chronology of events stretching from December 1978 to September 1981 that include dissemination of public information through the mass media, brochures, trade shows, newsletters and press releases; 15 meetings and open houses in northeastern Minnesota and the Twin Cities; 14 meetings with special interest groups and businesses; and 13 meetings with the Boundary Waters Canoe Area (BWCA) Citizens' Advisory Task Force. Appendix C summarizes public comment on the plan gathered from March–September 1981. Initial newsletters and news releases generated 500 specific comments within 65 letters, and the draft implementation plan induced 1,835 separate comments in 310 responses, with over two-thirds (214) of these responses coming from northeastern Minnesota.

The quantification and listing of public participation in these sections displays the dual sense in which public participation is made transparent in agency rhetorics. On one hand, it makes clear the variety of means by which publics could have been involved with the agency. On the other hand, it empties the substantive content of public participation and represents it instead in terms of aggregate data. This style of transparency fails to speak to the material concerns about management raised by citizens and interest groups, and gives the impression that the Forest Service is more concerned with the quantity of public participation generated than with the quality of ideas raised through participation.

The 1981 plan also puts great emphasis on the regulative framework in which management decisions are made. It goes to great lengths to show how the Forest Service has attended to all of the relevant laws and decision-making procedures in the development of the plan. For example, the first part of the plan responds point by point to the mandates of the Boundary

Waters Act, clearly displaying that the Forest Service has systematically proceeded through the relevant legislative mandate. The latter part of the plan offers management direction outside the scope of the Act, but it too links up the Forest Service's plan with prior policy moments, specifically, the 1974 BWCA Management Plan. This section discusses how to protect and rehabilitate resources and provide standards for management and public use of the area. Thus, the plan is clear in marking out its relationship with other plans, explaining on the first page how it fits into the planning hierarchy established by the National Forest Management Act (NFMA). This gives advocates an idea of alternative sites where policy decisions might be influenced. Overall, then, the plan makes public participation transparent and it displays its accountability to other governing institutions. But it spends little time addressing the concerns of citizens and interest groups, and misses an opportunity to develop legitimacy.

An example of this missed opportunity is the plan's treatment of the motorized portage issue, an issue that remains a point of contention twenty years after the publication of the plan. The issue of whether motorized portages violate wilderness values has provided a persistent source of conflict in the Boundary Waters controversy. Advocates have long disputed whether trucks and jeeps should be allowed as a means of transporting boats and gear across a small number of portages. Political and legal wrangling in the 1960s and 1970s ultimately led to the elimination of most motorized portages in the Boundary Waters Act of 1978. That Act also declared that three portages would be phased out unless, according to the statutory language, "the Secretary determines that there is no feasible nonmotorized means of transporting boats across the portages to reach the lakes previously served by the portages" (Public Law 95-495, Section 4[g]). The challenge of this passage lies in the word *feasible*. There has been no definitive resolution of what means of transportation are feasible, nor a clear sense of who is responsible for providing such a resolution.

The plan attempts to resolve the issue, however, in a way that ultimately ignored the public legitimacy of their decision. Attitudes about motorized portages provided a clear dividing line between many of the citizens and interest groups involved in the controversy (Gladden, 1990); but, instead of addressing these divergent opinions about public values—namely, the meaning and value of wilderness—the plan reverted to a technical, bureaucratic rationale for their decision. Appealing to the dictates of the Boundary Waters Act, the Forest Service frames its role as merely needing to answer the technical question of whether it is feasible to traverse the portages without motorized help. They answer this question in the plan by invoking the evidence of the Forest Service's attempt to portage a boat and gear across one of the portages that was traditionally motorized. By recounting this attempt in the management plan, the Forest Service relies on its institutional responsibility—namely, its need to appear accountable to the mandates of the Boundary Waters Act—to supercede questions about the soundness of the decision in relation to the attitudes and values of citizens and interest groups who have been party to the controversy.

Ultimately, the Forest Service executes its responsibility per the Wilderness Act without addressing the question of whether allowing truck portages constitutes environmentally responsible wilderness management in the eyes of the public. Although this allowed the Forest Service to displace the larger conflicts over the cultural value of wilderness to Congress, the failure to articulate this decision in terms of public values displays a missed opportunity for bolstering legitimacy of the agency's decision.

The 1986 Superior National Forest Management Plan

The 1986 Forest Plan (USDA-FS, 1986a) was part of the regular planning process for National Forests. Although the plan issues management directions for the entire Superior National Forest, it gives considerable attention to management of the Boundary Waters. Like the implementation plan discussed previously, the Forest Plan makes transparent both public participation and the Forest Service's relationship to the mandates of other governing institutions.

In terms of public participation, the 1986 plan consistently objectifies public participation. According to the Record of Decision, the Forest Service distributed over 1,100 Draft EIS and Plans, obtained comments from 213 individuals and federal, state, and local agency representatives, and made 13 changes during preparation of the final EIS in response to public comments. In quantifying this participation, the Forest Service again risks emptying participation of its content and obscuring the role that public comments played in the overall development of the plan.

Even where attempts are made to include public comments, these efforts work to minimize the quality and intensity of those comments. For example, the "Public Participation" section of the Record of Decision lists 13 changes made in response to public comments, but this section only contains that list without any mention of what the public comments were or how those comments affected these decisions. Also, in contrast to the 1981 plan, the 1986 Record of Decision (USDA-FS, 1986b) devotes considerable space to the reasons for its decision. Over half of the document is a section called "Reasons for the Decision" that intersperses paraphrases of public comments with agency responses. However, it is perhaps misleading to characterize these as public comments. First, as mentioned, these comments are actually paraphrases. Comments are never given by quotation but instead are summarized by the author of the document. Second, comments are never given in any depth but instead are reduced to one or two sentences. The text offers positions, but not the reasoning or support behind the positions. Third, comments on one issue are never articulated to comments on other issues as part of an overall perspective or position; instead, comments are separated by issue. These three characteristics contribute to the same rhetorical effect: we never hear an actual human voice in its depth or complexity. Although the comments may not be entirely transparent, their representation in the text makes those voices thin in quality and intensity. Audiences who did offer their comments are unlikely to find these representations adequate, thereby threatening the legitimacy of the agency and its decisions.

In addition, the 1986 plan highlights the service's close attention to Congressional mandates at the expense of pursuing public legitimacy. In this instance, the plan attempts to minimize the user conflicts within the Boundary Waters by showing that the Boundary Waters Act already moved many issues out of the realm of discussion. Because the Act specified that particular areas in the Wilderness would still be open to motor use, "the only decision that remains is what portions of the nonmotor area are to be managed for primitive or semi-primitive recreation opportunities" (USDA-FS, 1986a, p. 8). Thus, in affirming the mandate of the BW Act, the 1986 plan shuts down arguments about changes in motor usage and limits the scope of its inquiry to management within the nonmotorized areas. Here, the focus on a previous institutional mandate shifts attention away from ongoing user conflicts and keeps broader ecological arguments about motor usage out of the discussion. As a result, a pressing public issue gets ignored rather than addressed, suggesting that the agency is not responsive to public concerns.

The 1986 plan, then, provides a further example of the rhetorical processes that undermine the legitimacy of the agency. As public participation is made transparent and institutional concerns move to the foreground, the text offers little hope that agency personnel will give public voices substantial consideration. The 1993 plan offers further evidence (USDA-FS, 1993a).

THE 1993 BOUNDARY WATERS MANAGEMENT PLAN

Transparency of public participation is pervasive in the text of the 1993 plan. As Kathleen McAllister, Acting Superior National Forest (SNF) Supervisor, writes in the cover letter accompanying the plan, "Many [people] are deeply concerned about how the BWCAW is managed. More than 4,000 people commented in writing on the draft EIS and Plan" (USDA-FS, 1993a). In addition to mentioning the number of comments garnered, McAllister directly addresses citizens and gestures toward the effect that public participation has had in the decision-making process: "Thank you for sharing your concerns and ideas. Your feelings and information have helped me better understand the values this wilderness provides to America. Your ideas have helped me make a better decision." (USDA-FS, 1993a). Likewise, the Record of Decision especially amplifies public participation efforts. It states that in late 1990 and early 1991, open houses at nine sites in Minnesota and news releases sent to Midwestern media generated "over 1,100 pages of written responses from 620 individuals" (USDA-FS, 1993b, p. 12). Also, the Record of Decision lists a variety of factors involving public participation that influenced the service's decision: legislation, research, quality of experience, economics and public input. Each of these factors involved accounting for and ultimately responding to public concerns about the use and protection of the BWCAW. Finally, the Record has a separate section outlining "Public Involvement," which mentions the open houses, informal meetings, research reports on social conditions and visitor use in the BWCAW, task force meetings and solicitation of public comments. In all these instances, the repeated description focuses on public participation process more than content, thus minimizing the role of the content itself.

However, like the 1986 plan, the 1993 plan does attempt to connect public participation to actual changes in agency decision making. The EIS for both 1986 and 1993 contain a section entitled "Response to Public Comments" that contains an alphabetical list of all individuals who submitted written comments on the draft EIS, with numerical cross references to another part of the appendix describing the issue area and specific subject that their written communication addressed. These sections most closely link specific comments to specific changes, but as with the comments in the Record of Decision, the text fails to offer human voices in any depth or complexity. Also, the 1993 Record of Decision parallels the 1986 version of that document in the way it frames public participation. In 1993, the Record of Decision offers a series of "Reasons for the Decision" and lists four "Major Changes Made in Response to Public Comments." Again, these summaries do little to provide substantive representation of public participation; thus, they do little to adapt to the audiences who actually participated in the overall process.

In addition, the 1993 plan emphasizes the relationship between the Forest Service's decision and the mandates of other governing institutions. The first chapter of the plan, for example, summarizes the main laws and plans governing Boundary Waters management. These laws are also summarized in the Record of Decision, which contains a section entitled "Relationship With Other Laws." This section enumerates 14 laws that influenced the decisions contained in the plan and the environmental impact statement. It further explains how the decision is the direct result of a settlement agreement in 1988 that resolved an administrative appeal of the 1986 Superior National Forest Plan. As the Forest Service states in the summary of the EIS, the review and amendment of that Forest Plan is part of "a normal planning process, and it is an opportunity to improve our stewardship of the BWCAW based upon what we have learned since the Implementation Plan for the Act of 1978 was approved in 1981, and the Forest Plan was approved in 1986" (1993b, p. 1). Taken together with the representation of public participation in the plan, these characteristics of the text tend to minimize public participation in favor of the institutional rationales for decision making. Although the 1993 plan devotes several pages to public participation, like the other texts, it falls short in terms of generating legitimacy for the agency and its decisions. A brief look at criticisms of the 1993 plan reveals the failure of legitimacy.

FOREST SERVICE CRITICISMS AND THE EROSION OF LEGITIMACY

Even as the Forest Service attempts to respond to public comments in their plans, public response to management decisions suggests that the agency's rhetorical efforts have done little to bolster legitimacy in the eyes of the public. Various types of evidence support this claim. For starters, two appeals of the 1993 Management Plan reached U.S. district court and were decided in June 1997. It is interesting that the two cases argued against the plan from diametrically opposed positions. One case, brought by outfitters and three counties in the region, argued that the plan restricted access to the Boundary Waters and failed to consider the economic impact of the decision. The other case, brought primarily by conservation groups, challenged the plan's allowances for motorized use. Basically, one set of advocates claimed that the plan's regulations were too restrictive, while the other claimed that the regulations rolled back necessary restrictions. Ultimately, the district court judge heard and denied both suits. But even though this decision affirmed the Forest Service's position, the existence of the cases demonstrates that interested publics, from divergent perspectives, were not persuaded by the Forest Service's explanation of their 1993 decision. The Forest Service's arguments may have been justified institutionally as they were ratified by the courts, but those arguments did not seem to generate legitimacy among citizens and interest groups.

Further evidence of the erosion of legitimacy comes from comments made by advocates and legislators at oversight and committee hearings in 1995, 1996 and 1997. Again, the existence of these hearings and the proposal of legislation in themselves suggest discontent over Forest Service decisions. More specifically, advocates consistently raised the issue of public participation throughout the hearings as a reason for their disgruntlement. For example, in 1996, Senator Rod Grams put the following question to a Forest Service spokesperson: "You say you already have a public input mechanism of hearing from northern Minnesota. Why do we hear so many complaints from people in northern Minnesota who feel that they have been heard, but with a deaf ear? So in other words it is a kind of a charade of the process" (U.S. Senate, p. 47). Similarly, in 1995, Representative James Oberstar argued for a new citizens' management council for the Boundary Waters by appealing to the need for local residents to have a voice: "I hear it from the people here, and I am among the people and I know them—is you, the federal government, ask us for our comments. We tell you what we think, and then you go off and do something else. We need a way for people to have a voice" (U.S. Congress, p. 12). Todd Indehar, spokesperson for the group Conservationists with Common Sense, echoed Oberstar's argument: "We are asking for a partnership. We are asking for a voice. Obviously there is a national interest, and obviously the impact of federal land management policy falls disproportionately on those who live nearest the resource. . . . I think it follows that it is reasonable to have a partnership voice more than just an advisory and input voice; in other words, a management voice" (U.S. Congress, p. 96).

These comments clearly display the failure of regulative appeals as a means of generating legitimacy. Although the plans consistently emphasize how the Forest Service has followed the laws and regulations regarding public participation—by disseminating information and gathering significant, influential public input—citizens' concerns with voice suggest that the Forest Service has been unable to persuade relevant publics that their concerns have truly been heard. These comments in judicial and legislative settings, along with agency rhetorics, demonstrate that the agency's rhetorical efforts have not helped the Forest Service gain legitimacy in the eyes of the public.

This criticism, however, is not entirely the fault of Forest Service personnel. As the next section suggests, the agency operates under a set of legislative mandates that encourage a descriptive, formulaic inclusion of public comments in management plans. These mandates problematize Killingsworth and Palmer's criticism of the purpose of environmental impact statements. Although Killingsworth and Palmer criticize the defensive posture of EISs, the constraints on those documents make clear that the agencies must defend how a management plan fits into a larger regulative framework. Thus, in contrast to Killingsworth and Palmer, I argue that defense of agency action is a reasonable rhetorical purpose for the agency to pursue. To support this claim, I turn to a brief summary of these legislative mandates considered as rhetorical constraints.

RHETORICAL CONSTRAINTS ON THE FOREST SERVICE

In addition to the Wilderness Act of 1964 and the Boundary Waters Act of 1978, a variety of other environmental laws complicate Forest Service decision-making processes in the Boundary Waters. These laws emphasize public participation as a crucial part of that process. As public policy scholars Elise S. Jones and Cameron P. Taylor have argued, "the multitude of environmental laws enacted in the 1960s and 1970s opened up agency decision-making processes to public scrutiny and input, and provided citizens with greater opportunities to challenge agency decisions both administratively and in court" (1995, p. 311). In the case of Forest Service decisions about wilderness areas, the Multiple-Use and Sustained Yield Act (MUSYA), the Wilderness Act, the National Environmental Protection Act (NEPA) and the National Forest Management Act (NFMA) all help define the administrative decision-making process.

These acts articulate a structure of accountability for the Forest Service; that is, they enumerate ways in which the Forest Service must be responsive to various publics and other governing institutions in the course of policy formulation and implementation.

By articulating relationships of accountability for the Forest Service, these statutes also encourage the Forest Service to make certain kinds of arguments in their management plans, environmental impact statements, and records of decision. Because these laws require public participation, public participation becomes a central part of the agency documents. Further, management decisions themselves must be shown to be justified in relation to the dictates of other governing institutions. Thus, laws and regulations become important *topoi* in the construction of agency rhetorics.

Two statutes provide explicit evidence for this claim: the Wilderness Act and the National Forest Management Act. Both acts go into great detail about the relationship between the Forest Service and the public. The Wilderness Act, for example, explicitly addresses public involvement in the process of wilderness allocation and management because it mandates public notice of hearings and public participation in decisions. Section 3.a.1 requires the Secretary of Agriculture to maintain maps, legal descriptions, copies of regulations and public notices regarding pending additions, eliminations, or modifications of areas grandfathered into the National Wilderness Preservation System (NWPS). Section 3.d.1 makes more detailed provisions with regard to future inclusions into the system. This section calls for (1) giving public notice of proposed actions, including publication in the Federal Register and in local newspapers; (2) holding public hearings at locations near the proposed area; and (3) giving 30 days' notice about the hearings to governors, county boards, and concerned federal departments and agencies. Finally, section 3.d.2 requires that all views submitted under the previous provisions must be included with recommendations made to Congress or the President (Public Law 88-577). The Forest Service could be called to account for failing to adhere to any of these provisions in their decision-making process. Given these detailed provisions, it is not surprising that the texts they produce call special attention to public involvement in the decision-making process. Thus, the Wilderness Act clearly provides an impetus for these rhetorical characteristics in the management plans.

The National Forest Management Act and its accompanying regulations also encourage these characteristics. With regard to wilderness, the most significant part of NFMA is the detailed mandate for public input in preparation of the long-term land management and resource plans required for each unit of the Forest Service. The Act calls for public participation in the development, review, and revision of these plans, including holding "public meetings or comparable processes at locations that foster public participation in the review of such plans or revisions" (Public Law 94-588). Insofar as the Forest Service needs to show that they have met these demands, they are encouraged to make those practices transparent in their plans, and they are encouraged to emphasize opportunities for participation that have been made available to citizens and interest groups.

In addition, the NFMA encourages attention to institutional mandates, which produces the defensive posture that Killingsworth and Palmer criticize in their article. The NFMA states that Congress is supposed to hold the agency accountable for its decisions and actions. But in the process of holding the agencies accountable, the Act also encourages the Forest Service to defend its actions in relation to other institutions. This defensive posture is inevitable, given the framework of constitutional checks and balances set up by the Constitution. As Jones and Callaway suggest:

> Congressional oversight of federal agencies like the Forest Service is a desirable check and balance on the Executive Branch. In theory, as elected officials, members of Congress must be more accountable to the American people than are civil service agency personnel, and hence can help ensure that the agency manages the National Forests according to the public's priorities, whatever they may be. (1995, p. 342)

To the extent that Congress keeps the agencies in check, it is not surprising that the agencies produce arguments that defend their actions before a Congressional audience.

Thus, I would offer an alternative to Killingsworth and Palmer's criticism of the Forest Service's defensive rhetorical posture. Given the described rhetorical constraints, a defensive posture is reasonable, not itself a matter of critique. However, a more significant problem emerges when we see how this defensive posture dominates the documents and blocks possibilities for generating legitimacy. The rhetorical constraints provided by legislative mandates seem to encourage the Forest Service to generate a rhetoric rooted in a regulative view of institutions and legitimacy. These rhetorics crowd out the possibility of pursuing public legitimacy via other rhetorical means.

CONCLUSION

The ongoing controversies in the Boundary Waters—and in particular, criticisms of the role of public participation in making management decisions—suggest that the rhetorical conventions used in management plans to justify decisions often work to the detriment of the agency's interest and the public interest. They encourage an orientation toward justifying decisions within a regulative framework, rather than producing rhetorics that explain why a general public should perceive those decisions as legitimate. To use Burke's terms, the terministic screen employed in Forest Service management has produced a trained incapacity among agency officials for dealing with the issue of legitimacy (Burke, 1966).

The notions of terministic screen and trained incapacity help us characterize the institutional trap in which the Forest Service finds itself. Constrained to defend their decisions in relation to the mandates of other governing institutions, it becomes increasingly difficult for agency personnel to develop new rhetorics that might enhance the service's legitimacy. Peterson usefully characterizes how terministic screens can incapacitate us:

> Because our terministic screens constrain our observational possibilities to those in keeping with our occupational psychoses, we see new experiences in the terms provided by our past training. Of course, if the conditions of living have undergone radical changes since our terministic screen was developed, its serviceability may be impaired. Thus our training becomes an incapacity. (1997, p. 43)

For the Forest Service, the terministic screen articulated in legislative mandates has emphasized regulative concerns over public legitimacy in the production of public discourse. Although this terministic screen has helped the Forest Service focus attention on the necessity of public participation as part of the decision-making process, it has also incapacitated the agency, preventing it from representing that participation in ways that serve the end of public legitimacy. Public participation has come to be seen as an end in itself, something necessary to please other institutions, rather than a dynamic, ongoing activity that might help strengthen the legitimacy of the agency's decisions when represented in agency rhetorics. The lifeless lists of paraphrased comments and responses clearly shows that the rhetorical potential of public participation has gone untapped by the agency.

What, then, are the alternatives for new rhetorics of public participation in agency decision making? Rather than claim to have a definitive answer, I want to conclude by posing two somewhat extreme alternatives for developing new rhetorics of public participation that might help the agency bolster its legitimacy. I offer these not so much as practical solutions, but as radical speculations intended to draw out the practical and theoretical dimensions of the problem of generating legitimacy.

Killingsworth and Palmer hint at one possibility at the end of their chapter on the EIS when they claim that the "seeds of narrativity survive within the EIS process—in the practice of inviting the people of a region to respond to the EISs that affect them" (1992, p. 191). In a somewhat different vein, Peterson and Horton (1995) argue for a more complex public discourse when she asserts, "Because environmental policy is public policy, its discourse should throb with the sometimes contentious sounds of public debate" (p. 140). Both of these comments suggest that the divergent voices generated through public participation must be amplified so that mutually agreeable and beneficial courses of action can be identified. Indeed, a management plan that presents competing narratives in their own voice, that recognizes distinct positions and groups in their complexity, represents an ideal that may be unreachable but may also be worth reaching for.

Management plans that let these competing narratives speak in their distinctive voices would foreground the normative dimension of institutional action and highlight normative grounds for legitimacy by focusing on the values and norms of various interest groups. The legitimacy of agency action would be produced not through the assertion of a set of allegedly universal values, however; it would emerge from the creative identification of points of affiliation between competing groups and their often competing values. The possibility of such an identification is displayed in Peterson and Horton's essay, and shows the practical contribution that rhetorical criticism can make to the resolution of environmental disputes.

At the same time, the practicality of this proposal is questionable because it suggests nothing less than the decentering of agency decision making. Rather than have management decisions emanate from the centered subject of the Forest Service, policies would include multiple voices in the presentation of the plan—both in written documents and in public displays or presentations of the plan. The Forest Service likely would see such a proposal as a threat to its power. However, the release of an agency plan under these conditions could diminish the sense that the plan is a document produced by a monolithic, federal voice. These texts would make dialogue a concrete textual reality rather than merely a philosophical ideal, and provide a more palpable sense that public voices had contributed to the production of public policy.

This proposal raises an additional practical question in the context of this chapter and leads us to a second rhetorical alternative. How would the inclusion of more voices help the Forest Service address the problem of legitimacy? Indeed, it might not be beneficial for the Forest Service's distinctive voice to recede in their own documents. So, a second, somewhat heretical alternative: perhaps the Forest Service could be better served by explicitly stating, prior to the participation process, the agency's best thinking on an issue. In place of a de-centered dialogue, the agency would produce texts of explicit advocacy like other groups and organizations. This alternative could help the Forest Service get out from under the cloud of mistrust and suspicion about the motives behind its public participation efforts by encouraging the agency to be up front about its preferences and positions.

This option, too, raises significant obstacles, especially because it suggests a radically different role for public agencies. It also suggests a different participation process: one in which the Forest Service generates advocacy at the beginning of the process. The agency would sit at the table as one more player in the process with other groups, rather than the puppet master who seems to control the process for its own ends. Further, it would make the issue of legitimacy a consistently salient issue in the agency's production of rhetorical documents. In contrast to the first alternative, the goal of the agency's texts would not be to enact dialogue internally, but to promote dialogue externally among different groups that share a stake or interest in the decision. Instead of trying to create a textual form that enacts an ideal state of affairs, the agency would need to consider how their texts might contribute to an improved state of affairs.

As with the first option, the practical problems of such a radical shift make this alternative unworkable. However, contemplating this alternative would remind agency personnel that they are not above the fray but are an active participant in processes of public advocacy. Although they may attempt to accommodate multiple interests and values in management, their rhetoric is never going to be perceived as neutral by audiences interested in their decisions. Thus, we return to the underlying assumption of this chapter—that the Forest Service can benefit from closer attention to the rhetorical implications of their discourse. No matter how the agency might deal with the issue of legitimacy, it can only do so effectively through a rhetorical lens.

CHAPTER 8

PUBLIC INVOLVEMENT, CIVIC DISCOVERY, AND THE FORMATION OF ENVIRONMENTAL POLICY: A COMPARATIVE ANALYSIS OF THE FERNALD CITIZENS TASK FORCE AND THE FERNALD HEALTH EFFECTS SUBCOMMITTEE

STEPHEN P. DEPOE

A growing number of scholars are examining the structures and processes of public involvement in the formation of environmental risk policy (Belsten, 1996; Kasperson, 1986; Katz & Miller, 1996) and the resolution of environmental management disputes (Daniels & Walker, 1996; Daniels, Walker, Carroll, & Blatner, 1996; Walker, Daniels, Blatner, & Carroll, 1996). Emerging from this literature is the axiom that different processes and mechanisms for citizen involvement reflect different assumptions about risk, rationality, and human judgment (Fiorino, 1989b; Juanillo & Scherer, 1995; Plough & Krimsky, 1987; Waddell, 1996).

When government and industry operate from a traditional approach to risk communication, resultant policy decisions often do not win sustained support of affected communities. The traditional approach to risk communication employs a technocratic model, which excludes public input into decision making altogether; or a one-way Jeffersonian model, which involves efforts to control public opinion through a unilateral transfer of expert knowledge (Waddell, 1996). Public participation mechanisms grounded in a traditional approach feature the one-way transmission of messages from monolithic institutional sources, a decide-announce-defend strategy for policymaking, and limited public access to officially recognized communication channels and forums (Fiorino, 1989b, 1990). A good example of a participation mechanism grounded in a traditional approach to risk communication is a standard public hearing (Katz & Miller, 1996).

An increasingly popular alternative to the public hearing is the *citizen review panel*, a participatory structure that allows for selected community members to interact with scientists and technical experts, to evaluate evidence and debate alternatives, and to make environmental and health policy recommendations to government authorities (Fiorino, 1990). Citizen advisory groups may be authorized through legislative or regulatory mandates (such as the Federal Advisory Committee Act), or be formed through voluntary cooperation among stakeholder groups.

Citizen review panels would seem to offer increased opportunities for public involvement based on more two-way interactions between technical experts and the broader public (Waddell, 1996). However, such panels are not always successful in promoting effective community collaboration. When implemented based on conventional notions about the superiority of expert knowledge, or without legitimate power to influence policy, citizen review panels can serve more of a co-optation or ritualistic function than as a source of community empowerment (Juanillo & Scherer, 1995). The mere creation of an outlet for community opinion or advice does not guarantee that an environmental policy process will produce publicly legitimated decisions (Rowan, 1994).

What is needed is a clearer set of criteria for evaluating how well various public involvement mechanisms empower those who have a stake in environmental policy decisions. Juanillo & Scherer (1995) have moved in this direction by articulating an alternative to traditional risk communication that they have termed a *dialectical* approach. A public participation mechanism grounded in this more democratic approach would feature the following characteristics (Fiorino, 1990; Belsten, 1996):

- Encourages direct involvement of citizens who represent diverse perspectives;
- Provides for adequate access to information channels and resources;
- Provides for frequent face-to-face discussions among stakeholders over time;

- Institutes an equitable distribution of power in collective decision making.

A dialectical approach to risk communication should promote *civic discovery*, defined by Walker, Daniels, Blatner, and Carroll (1996) as the authentic and mutual engagement by communities, industries, and government agencies in conflict resolution through collaborative learning. Civic discovery, which "features dialogue, constructive argument, and negotiation: key elements of public deliberation," can yield decisions that are both technically sound and more likely to be accepted by affected communities (p. 6).

Over the past decade two citizen review panels have addressed environmental and human health issues associated with the Fernald Environmental Management Project (FEMP), a former Department of Energy (DOE) nuclear weapons production installation in southwest Ohio that is now undergoing environmental remediation as a Superfund site. These two citizen advisory groups, the Fernald Citizens Task Force (FCTF) and the Fernald Health Effects Subcommittee (FHES), are the focus of this study.

The study offers an analysis of both advisory groups, based on direct observation of group meetings, along with data found in meeting minutes, transcripts, and reports generated by the groups. The principal analysis covers the time period from 1993 to 1997, with an epilogue that summarizes the groups' activities since 1997. Through comparing and contrasting the structures and discursive practices of these two citizen advisory groups, this study assesses the degree to which the FCTF and FHES fulfill the participatory ideal of civic discovery.

THE FERNALD ENVIRONMENTAL MANAGEMENT PROJECT (FEMP)

Fernald is a former uranium processing plant owned by the Department of Energy (DOE). The site, located 18 miles northwest of Cincinnati, Ohio, was part of the U.S. nuclear weapons complex that processed uranium ore and produced uranium metal products to be used in the manufacture of nuclear weapons at other sites across the country from 1953 to 1989 (U.S. Department of Energy [DOE], 1995b). In 1989, the facility ceased production after radioactive environmental pollutants such as uranium were found in nearby land and water and it was placed on the U.S. EPA's National Priorities List for Superfund. Since 1991, the facility has been undergoing environmental remediation directed by an agreement between the Environmental

Protection Agency (EPA) and the DOE according to federal guidelines mandated in the Comprehensive Environmental Response, Compensation, and Liability Act (CERCLA) (US DOE-FEMP, 1994).

Production and disposal practices at the Fernald site have caused contamination of the air, water, and soil on site and the surrounding area from uranium and other hazardous radioactive materials (Fernald Environmental Restoration Management Corporation [FERMCO], 1995). The total waste and contaminated material that needs to be remediated at the Fernald site is over three million cubic yards (Fernald Citizens Task Force [FCTF], 1995). DOE and several contractors have been charged with the management of the Fernald site over its history. Most recently, the Fernald Environmental Restoration Management Corporation (FERMCO), a subsidiary of Fluor Daniel, assumed responsibility for the cleanup of the Fernald facility. The FERMCO name was changed to Fluor Daniel Fernald, Inc. (FDF) in September, 1996.

Information about environmental pollutants released by Fernald production activities were kept secret from the public for most of the thirty-eight years the facility was in operation. In the mid-1980s, when contamination was found in nearby drinking water wells, public awareness and concern about environmental contamination increased, and the citizens began urging government agencies to clean up the site. As environmental remediation activities commenced, area residents remained highly skeptical of DOE and government commitments to clean up the site. Beginning in the mid-1990s, DOE and its contractor, FDF, implemented programs such as quarterly public meetings, special workshops, and stakeholder briefings mandated by CERCLA to involve citizens in the Fernald cleanup process. Over time, citizens living near Fernald have become active and informed participants in cleanup discussions; however, the interaction between site officials and Fernald citizens has continued to be influenced by the culture of secrecy that was dominant throughout the nuclear weapons complex during production years (Kinsella, 2001; Kuletz, 1998; Taylor, 1998; US DOE, 1995).

THE FORMATION OF THE FERNALD CITIZEN ADVISORY GROUPS

A number of historical developments led to the formation of the Fernald Citizens Task Force (FCTF) and the Fernald Health Effects Subcommittee (FHES). The extensive environmental damage produced by DOE production of nuclear weapons materials affected many thousands of American citizens across the country and created a significant environmental policy challenge for the federal government. As the magnitude of the environmental legacy of the Cold War became clearer, so did voices calling for more citizen involvement in the determination of DOE's environmental remediation priorities. In 1993, the Federal Facilities Environmental Restoration Dialogue Committee (FFERDC) issued a report that recognized that "individuals affected by environmental remediation activities were not being given sufficient opportunity for meaningful dialogue or to provide input regarding the remediation process" (FCTF, 1995, p. 12, citing Keystone Center, 1993).

As a result of this report and growing public pressure, in 1993 DOE began to form site-specific advisory boards (SSAB) at a number of installations within the nuclear weapons complex. At that time, the DOE, U.S. EPA, and Ohio EPA (OEPA) commissioned the Fernald Citizens Task Force as the Fernald site-specific advisory board to provide a community perspective and develop a public consensus about Fernald environmental remediation through focused and informed stakeholder input on cleanup decisions.

The Fernald Citizens Task Force decided to make a limited number of recommendations to Fernald decision makers. In July 1995, the FCTF issued recommendations on remediation levels for the site, methods of waste disposition, priorities for remediation, and future use of the land now occupied by the facility. The task force has continued to meet on a regular basis to advise DOE on remediation options and priorities.

At the same time that site-specific advisory boards were forming at Fernald and elsewhere in the DOE complex, community members living near these installations began to express a growing level of concern about risks to human health posed by past and present production and remediation activities. In particular, citizens wanted to know about potential health impacts of hazardous chemical and radioactive materials that had been released into the surrounding air, water, and soil. In response to that concern, the Centers for Disease Control and Prevention (CDC) and other government agencies initiated studies at several DOE installations in an attempt to identify the types and amounts of hazardous materials released into the environment over time and to characterize the potential risks to the health of installation workers and neighboring residents.

An additional step was taken by the federal government in 1994 when, as part of an agreement between DOE and two agencies within the Department of Health and Human Services, the Centers for Disease Control (CDC) and Agency for Toxic Substances and Disease Registry (ATSDR), a number of citizen advisory groups were authorized to solicit community input concerning the scope and direction of federal health research and other public health activities connected with past or continuing operations at Fernald and other DOE sites. The Fernald Health Effects Subcommittee (FHES), established in 1996, was the sixth site-specific advisory board created to make recommendations on collection of environmental health data near DOE weapons production facilities. The FHES began holding quarterly meetings and was tasked with producing recommendations concerning the need for additional community health research, including a site-wide epidemiological study, at Fernald.

Now that a context has been provided for the formation of the two Fernald citizen advisory groups, I turn to an analysis of each group through the application of the features of the dialectical model of risk communication.

FCTF VERSUS FHES, 1993–1997: A COMPARISON OF STRUCTURAL ELEMENTS

This section examines the structural and operational elements of the Fernald Citizens Task Force and the Fernald Health Effects Subcommittee, highlighting similarities. A discussion of each major element of the comparative analysis is followed by a summary table.

DIRECT INVOLVEMENT OF DIVERSE PERSPECTIVES

FCTF has been characterized by direct participation of interested and informed citizens who represent diverse perspectives. FCTF members were chosen by Eula Bingham, an independent convener from the University of Cincinnati. Professor Bingham enjoyed significant credibility in the community and was knowledgeable about local stakeholder issues. She selected 14 citizens representing multiple views in the Fernald area. The FCTF included Fernald-area residents, business persons, academics, local government officials, health professionals, labor representatives, teachers, and members of a local citizens activist group, the Fernald Residents for Environmental Safety and Health (FRESH). The citizens were joined by ex-officio members from DOE, US EPA, OEPA, and FERMCO. FCTF participation in site decisions was much more involved than that of citizens who might attend a meeting or workshop and provide input by writing a letter or calling a site official. Because the FCTF was designed to be an official voice for the Fernald public, task force members were allowed to make recommendations directly to Fernald decision makers.

The membership of the Fernald Health Effects Subcommittee was both more demographically diverse and less immediately affected by site activities than the membership of the Fernald Citizens Task Force. Fernald-area residents composed less than half of the committee, with the other half hailing from other communities in the greater Cincinnati area. Two African American women served on the committee, as did four doctors, an engineer, and two labor representatives. The high number of doctors and other professionals on the committee reflected the traditional concerns of the Centers for Disease Control (CDC), the agency selecting the panel, about the highly technical nature of the information to be discussed.

Perhaps the most important aspect of each panel's membership was its chairperson. John Applegate, an environmental law professor with significant mediation experience and an interest in public participation and environmental policy, was selected as the inaugural chair of the FCTF. After becoming chair, Applegate became active in national-level DOE activities, including service on DOE's Environmental Management Advisory Board (EMAB), and also wrote extensively on the FCTF decision-making process (Applegate & Sarno, 1996, 1997; Applegate, 1998). His leadership and facilitation skills were quite strong, and his patient guidance helped FCTF to develop a clear focus and set of recommendations.

Joseph Farrell, chair of the FHES, was a less inspired choice. Trained as researcher in the area of wastewater sludge, he retired from a career as a University of Cincinnati professor and consultant for the EPA prior to accepting his new position. From the inception of the FHES, he served as a more passive, bureaucratically minded chair, often hurrying through segments of the agenda dedicated to public comments and discussion to leave enough time for the next technical presentation. He demonstrated little empathy with the Fernald-area residents, who consistently expressed significant concerns about risks Fernald production may have posed to their health.

Table 8.1 summarizes the differences between the FCTF and FHES concerning involvement of diverse perspectives.

Access to Information Channels

From the outset, FCTF members were provided with sufficient access to information channels and resources. DOE provided the FCTF with a consultant to give independent, technical support on a regular basis. When the FCTF was formed, DOE decisions about Fernald site cleanup were already

Table 8.1
FCTF vs. FHES: Direct Participation of Diverse Perspectives

FCTF	Structural Element	FHES
• Members chosen by independent convener • 14 members representing multiple views in Fernald area • Members chosen were already actively involved in site activities • Group joined by ex-officio members (DOE, EPA) • Chair was environmental law professor with mediation experience	Encourage direct involvement of citizens representing diverse perspectives	• Members chosen by CDC • 15 members representing diverse views in Cincinnati area • Wide range of knowledge and background in Fernald issues • Group joined by ex-officio members (Ohio EPA, Ohio Department of Health) • Chair was retired environmental engineer

in progress and DOE was approaching a Record of Decision (ROD) on several of the operable units on site. FCTF members had to catch up on enormous amounts of technical information quickly if they were to develop recommendations in time to influence DOE decisions. To facilitate this, the technical consultant developed a "tool box" of information, presented in a basic and understandable format, for the group to study. The consultant also verified DOE information given to the FCTF to develop trust in the information base for decisions. FCTF members were given immediate access to key Fernald site personnel to answer questions or provide additional information. As a result, the FCTF began to ask for information that had not yet been generated for site use and were therefore able to influence the site agenda through their information requests.

The CDC did a less effective job in providing the members of the Fernald Health Effects Subcommittee with adequate and usable information. During each of the group's first five two-day meetings, a significant number of hours were devoted to long, ponderous presentations of technical information, made by a host of government officials and experts, about topics ranging from principles of toxicology and epidemiology to the results of a major-dose reconstruction study to the history of the site. Little effort was made to adapt the information to the understandings of committee or to those of the broader audience attending the meetings.

Copies of presentation transparencies and accompanying bibliographic material were made available with little in the way of technical support or summary. All of these technical presentations followed a traditional, one-way information flow, with questions coming at the end of the allotted time period. CDC tried to supplement their informational efforts by offering subcommittee members an extension course in epidemiology offered on Tuesday afternoons via satellite from the University of North Carolina. Two committee members took the course. CDC and the FHES chair showed uneven skills in responding to the questions or requests for more information made by FHES members. For instance, at the May 1997 meeting, an extensive (and largely tangential) presentation was made to the group on the hazards of radon in the home, while the request made by another member for a discussion of risks posed to drinking water by the site was neglected. The problem faced by the FHES was information overload with little adaptation made to members' needs or interests. In sum, the CDC approach to conveying information followed a traditional, technocratic model of risk communication.

Table 8.2 summarizes contracts between the two advisory boards in the area of access to information.

Table 8.2
FCTF vs. FHES: Access to Information

FCTF	Structural Element	FHES
• DOE supplied consultant to provide independent technical support • Consultant developed "tool box" • FDF and DOE activities influenced by members' questions	Provide sufficient access to information channels and resources	• CDC, NCEH, NIOSH, ATSDR provided steady dose of informational presentations • CDC provided committee members and audience with copies of transparencies and other technical information • CDC provided extension course for members • Mixed success shown in responding to member requests

FACE-TO-FACE DISCUSSION OVER TIME

FCTF members engaged in consistent face-to-face discussion over time. The FCTF met monthly during its first two years of existence, and has met on a near-monthly basis ever since. Meetings were usually held on Saturday mornings, and typically lasted about three hours. In 1996, the FCTF formed issue-based committees to continue the advisory work. In addition, during the formative first few years, group members met informally at social gatherings to get to know each other better. Stable membership and regular meetings allowed task force members to develop relationships, discover common values, and influence each other, all of which are important parts of civic discovery. The regular meeting schedule also helped to maintain a decision time frame for the group, including time for members to study and deliberate on technical materials prior to making decisions. Task force members and staff also kept other community members informed about activities and pending decisions through the use of both formal (fact sheets, reports) and informal (person-to-person, small group) communication channels.

In convening the FHES, the CDC adopted a different strategy: fewer meetings of longer duration. The group met quarterly, with each meeting lasting almost two full days. Very little formal or informal activity took place in between scheduled meetings. The group struggled to find a meeting time and place convenient for group members. The initial decision to hold meetings in a northern Cincinnati suburb several miles from the Fernald community angered Fernald community members on the committee. Holding the bulk of the meeting sessions during the workday created significant hardship for some of the members. One member (Pam Dunn, a member of FRESH and the appointed liaison between FHES and FCTF) did not attend the first several meetings because of work conflicts. The CDC also did a poor job coordinating FHES activities with scheduled meetings of other Fernald-area groups. On two occasions, evening sessions designed to elicit community input and comments were scheduled at the same time as previously arranged community meetings with DOE officials.

The FHES meetings themselves did not encourage direct interaction among committee members, or between the committee and the usually small audience in attendance. Meeting agendas, procedures, and even the arrangement of tables and chairs in the room created physical and symbolic distance among the meeting participants. The agendas typically called for limited time for questions from the audience, usually after a number of technical presentations. FHES members were asked to address the chair, in formal parliamentary fashion, and not each other. Even the tables where the FHES members were seated were arranged so that the members had their backs to the audience. Each meeting resembled a combination of a formal hearing and a scientific lecture.

Some of the formal features of the FHES meetings were altered in a positive way during the second afternoon of the May 1997 meeting. Based on FHES members' suggestions, chairs for the audience were moved closer to the committee members' tables and placed directly behind the members, thereby increasing eye contact and decreasing distance. A full two hours were devoted to a general roundtable discussion in which committee members asked a wide range of questions to a panel of government officials and scientists. The interaction during that session was a highlight of the early history of the FHES.

Table 8.3 highlights differences in face-to-face interaction.

Equitable Distribution of Power

During its initial period of existence, the FCTF operated based on a relatively equitable distribution of power among its members and between the advisory group and DOE. The FCTF was formed based on a democratic model with equal voting power within the group. It set its own agenda, agreeing to provide input on four issues concerning Fernald site cleanup:

Table 8.3
FCTF vs. FHES: Person-to-Person Discussions

FCTF	Structural Element	FHES
• Met monthly since 1993 • Engaged in formal and informal gatherings • gave themselves time to deliberate • disseminated results throughout community	Provide for frequent face-to-face discussion over time	• Met quarterly since June 1996 • Chose inconvenient meeting times and places • Did not coordinate meeting times with other Fernald meetings • Formal gatherings, resembling public hearings • some progress made in format at last meeting

remediation levels, waste disposition, remediation priorities, and future land use. Task force members identified these four as the problems to be addressed by their group. Part of identifying the problems was developing an 18-month work plan organized around the time frame for DOE technical milestones so that recommendations would be made in enough time to influence site decisions. All of these moves increased the likelihood that the FCTF would be influential.

As representatives of the entire Fernald community, FCTF members considered community opinions outside of the task force to be as important as their own. The FCTF demonstrated this by holding community workshops concerning controversial recommendations in an attempt to obtain complete community input. All FCTF meetings were open to the public and several workshops and presentations were held to elicit broader community perspectives than the FCTF.

To promote civic discovery, a level playing field should be established between citizens and government officials so that citizens are allowed to define issues and question the experts (Walker et al., 1996). The FCTF was empowered to question technical experts and dispute evidence. The equal level was also evident in the FCTF charge to make recommendations to DOE and EPA and in the reality that FCTF recommendations were incorporated into the CERCLA process.

In its first few years of existence, the FHES had less success in developing a clear agenda or a definite role in government health policy concerning the Fernald site. According to the draft mission of the FHES, the subcommittee was established "to provide CDC and ATSDR with the community's advice on their health research and public health activities connected with past or continuing operations at the Department of Energy's Fernald site" (FHES, 1996, n.p.). The range of policy options available to CDC and ATSDR appeared to be somewhat limited due to funding and the retrospective nature of health research. The major decision, whether to fund a full community-wide epidemiology study at Fernald, was an important one, but no time frame for that decision was set, nor was the extent or role of the FHES in making that decision clearly stipulated. In fact, FHES members held very few votes, since almost all of the time in each meeting was devoted to informational presentations. Decisions concerning agenda items were left to Dr. Farrell, the chairperson, and Steve Adams, CDC representative on the FHES.

The FHES did a poor job of soliciting community opinions on Fernald health issues. The meetings generally drew little media attention, and atten-

Table 8.4
FCTF vs. FHES: Distribution of Power

FCTF	Structural Element	FHES
• Group identified and prioritized tasks • Group developed meaningful decision timetable • Used democratic model with equal power among members • Established consensus values before making recommendations • Considered other community opinions • Presented recommendations that were taken seriously by DOE	Institute an equitable distribution of power in collective decision making	• Had unclear mandate • Did not establish decision timetable • Voted on very few matters; chair and CDC facilitator decide next steps • Did not gather or heed other community opinions • Had limited resources to gain information or influence

dance by members of the community was sparse due to time and scheduling conflicts. FHES members were provided with copies of comment cards filled out by community members who attended the meetings or who sent inquiries to CDC, but the chair, Joseph Farrell, chose to pay little attention to those comments in subsequent meetings. Sadly, neither the CDC nor the FHES leadership seemed to possess the resources or will to expand the visibility or influence of the group.

Table 8.4 summarizes how each board fared with respect to distribution of power.

SUMMARIZING RELATIVE EFFECTIVENESS, 1993–1997

After considering the concerns of fellow community members, the FCTF in 1995 made the final recommendation for on-site storage with only one dissenting member. Of a total of over 3 million cubic yards of contaminated materials, it was recommended that 800,000 cubic yards of high-level waste be sent to EnviroCare in Utah by railroad transport and

2.4 million cubic yards of low-level waste be stored on the Fernald site in an above-ground disposal cell.

The task force decided that although moving Fernald's high-level hazardous waste off site was essential to protect the Great Miami Aquifer that lies directly beneath the site, it was reasonable to store the low-level waste on site. The final recommendation was submitted by the FCTF chair to DOE office of environmental management in July, 1995. This difficult decision was the product of civic discovery, developed and nurtured over a five-year process in which the FCTF gained credibility as a representative body of stakeholders led by an excellent chair, obtained sufficient informational resources in usable portions from a technical consultant provided by DOE, engaged in regular face-to-face meetings over time to generate trust among the members and between members and DOE officials, and developed equable power arrangements within the group and with DOE and regulatory agencies. These developments were due to good people working within a dialectical risk communication framework.

In contrast, the FHES faced significant risk challenges in its efforts both to learn about and to advise policymakers about the investigation and reporting of human health risks associated with past and present activities at the Fernald site. Many citizens living near Fernald lacked the technical training necessary to understand the methods and results of health risk studies. As a result, they became frustrated with time-consuming and costly research that seemingly failed to provide conclusive answers to questions about potential threats to their health. Additionally, community outrage was compounded when those charged with conducting health risk research and communicating results to the public demonstrated an inadequate understanding of legitimate community needs and interests.

The CDC did little to support the FHES. Its membership selection, particularly the selection of chairperson, left something to be desired. CDC and its sister agencies inundated FHES members with technical information in the typical fashion of traditional risk communication, but did little to provide guidance for what the group should do with the information. CDC scheduled meetings were both too lengthy and too infrequent, with little to no support between meetings. Meetings were often held at inconvenient times and places, in conflict with other community events, and with little effort made to promote broader community input or participation. The role of FHES in helping CDC make the decision about whether or not to conduct a major epidemiological study at the site was never clarified. In many ways, the first two years of the Fernald Health Effects Subcommittee represented just about everything that could go wrong with a citizen advisory panel mechanism that relies on conventional wisdom about risk communication and the role of the public in environmental decision making (Katz & Miller, 1996). Its ineptness in promoting civic discovery offers lessons every bit as important as those provided by the spectacular success of the Fernald Citizens Task Force.

Epilogue: Fernald Citizen Advisory Panels Since 1997

Since 1997, Fernald's two citizen advisory panels have continued to follow trajectories suggested in the above analysis. The Fernald Citizens Task Force changed its name to the Fernald Citizens Advisory Board (FCAB), and underwent a transition in leadership in 1998 when area teacher James Bierer replaced John Applegate as the chairperson. Other than those changes, however, the board has operated in much the same manner as it did between 1993 and 1997. FCAB meetings and related activities have been the product of direct citizen involvement, adequate access to information, and ongoing face-to-face interaction among stakeholders and responsible parties, all of which have generated for the FCAB a significant amount of voice in and influence on site cleanup decisions.

The FCAB has continued to serve the community well by focusing on the development of consensus-based recommendations on site cleanup issues. Specifically, since 1997 the FCAB has issued recommendations on natural-resource restoration on site, cultural management issues, waste disposition and transportation, budget priorities, and remediation options for a number of site cleanup projects (Fernald Citizens Advisory Board, 2002).

Beginning in 1999, the FCAB embarked on a "Future of Fernald" planning process involving a series of community workshops at which FCAB members, in conjunction with other area environmental and historic-preservation groups, facilitated discussions concerning how the Fernald site may be transformed into a community asset after remediation is completed. Discussions have included the possible development of a multiuse educational facility housing environmental education, Cold War history, and Native American artifacts, along with official documents generated as part of the administrative record of Fernald's cleanup activities. The Future of Fernald process, which was ongoing as of the end of 2002, has involved over 100 area residents and other interested parties. The process has been instrumental to the formation by FCAB of a "stakeholder vision for the future of

Fernald," and holds significant promise for yielding additional community-generated recommendations for DOE as well as for state and federal regulators (Goetz, 2000). The FCAB's Future of Fernald project serves as an additional example of a public participation effort aimed at producing a sense of civic discovery (Sarno, 2001).

The Fernald Health Effects Subcommittee has been much less successful in producing robust and meaningful public participation in the area of health research. In the years since 1997, CDC facilitators and FHES leadership made some attempts to improve meeting formats to increase the quality of interaction among board members as well as the amount of community involvement. More meeting time was devoted to roundtable discussions among members and between members and CDC and the National Institute for Occupational Safety and Health (NIOSH) researchers making presentations. The format was also changed to allow for one evening session for the general public during each quarterly meeting. These changes produced a small degree of improvement in the perceived productivity of the meetings.

A number of significant health studies were presented to the FHES for comment, and were ultimately released for publication. A dose-reconstruction study completed by John Till examined potential risks of contracting lung cancer based on different hypothetical scenarios involving residential exposures to radon gas from silos on site containing radium and other mixed waste (RAC, 1998). The study, which was peer reviewed by the National Research Council (NRC), produced findings with a high degree of uncertainty, which only fueled the controversy over the need for additional research (NRC, 1997). The CDC funded two additional risk assessment studies that attempted to estimate the extent to which Fernald-area residents may have experienced increased mortality rates from lung cancer and other cancers due to exposures to radioactive materials during Fernald's production years of 1951–1988 (Devine, Qualters, Morrissey, & Wall, 1998, 2000).

The relatively small amount of excess risk suggested by these studies led CDC to conclude that a more extensive epidemiological research project at Fernald was not warranted (Melcer, 1999). Based on that conclusion, the CDC further decided in 2001 to terminate the Fernald Health Effects Subcommittee. This decision was made despite significant community dissent expressed at the final FHES meetings and elsewhere. Many FHES members and other Fernald-area residents felt both that the CDC decision was not justified on substantive grounds, and was also made without sufficient consultation with the affected community (Bonfield, 2001a, 2001b). For many in the community, the CDC decision to terminate the FHES provided final confirmation of the inadequate degree of democratic participation that characterized the history of the ill-fated health advisory group.

The future of other DOE-related health effects advisory groups is under significant scrutiny. In 1999, CDC commissioned the COSMOS Corporation to conduct an evaluation of the health effects subcommittee advisory process. Two years later, COSMOS issued a report containing 17 recommendations concerning ways that the structure and operations of the subcommittees can be strengthened (COSMOS, 2001a, 2001b, 2001c). The recommendations include ways of improving:

- How the health effects subcommittees provide and submit advice to government agencies and how the agencies respond to that advice; and
- The advisory system's efforts to increase credibility in public health activities and research to promote trust between citizens and federal agencies, and efforts to obtain a broad range of public involvement in public health activities and research at DOE sites.

One can only hope that as the DOE, CDC, and other federal agencies seek to improve public participation processes, they review the relative success of the Fernald Citizens Advisory Board and the Fernald Health Effects Subcommittee.

CHAPTER 9

PUBLIC PARTICIPATION OR STAKEHOLDER FRUSTRATION: AN ANALYSIS OF CONSENSUS-BASED PARTICIPATION IN THE GEORGIA PORTS AUTHORITY'S STAKEHOLDER EVALUATION GROUP

CAITLIN WILLS TOKER

Over the past few decades, the environment has become a topic of considerable debate (Dahlberg, Soroos, Feraru, Harf, & Trout, 1985; Kraft, 1999; Shabecoff, 1993). Chief among the concerns in this arena is the role of citizens in environmental policy decisions (Bacow & Wheeler, 1984; Krimsky, 1992; Trumbo, 2000). Participation in environmental decision making has been a part of the national agenda since Nixon outlined a role for the public in the 1969 National Environmental Policy Act [NEPA]. This law mandated that agencies complete an elaborate process of study and review whenever their proposed action affected the environment. Citizens were given the opportunity to comment at various stages of the process through public hearings and comment periods. Since then, increased environmental consciousness has produced a proliferation of federal and state laws with similar provisions. Yet, traditional participation mechanisms have done little to facilitate the type of collaboration originally intended (Fiorino, 1990; Vaughan & Seifert, 1992). Community members often either fail to participate due to a felt lack of knowledge or they become frustrated because these processes appear to be little more than public relations efforts employed to convince them to accept decisions already made (Katz & Miller, 1996).

In recent years, researchers have begun promoting more direct forms of participation and have identified a consensus-based stakeholder approach as the solution to this problem (John & Mlay, 1999; Juanillo & Scherer, 1995; Renn, 1992). Although "there is no single description in the literature of these new approaches to resolving environmental disputes" (Crowfoot & Wondolleck, 1990), they all follow a model whereby rational, authoritative consensus decisions are reached through the free and open deliberation of representative and equal stakeholders. This model reflects the deliberative approach to political disagreement advocated by democratic theorists in recent years. Although a number of case studies point to the difficulty and even the impossibility of securing this ideal in practice (Murdock & Sexton, 1999; Spyke, 1999; Waddell, 1996), many researchers refuse to critique its use. Instead, they attribute the failure of such efforts to the incorrect implementation of the model rather than to flaws in the model itself (Katz & Miller, 1996).

The Georgia Ports Authority's (GPA) decision to create the consensus-based Stakeholder Evaluation Group (SEG) to deal with opposition to its proposed deepening project reflects this trend. The GPA announced its plans to deepen the sandy and naturally shallow Savannah River in their Tier I Draft Feasibility and Environmental Impact Study issued in May 1998. Although the river was just deepened to 42 feet in 1994, the GPA reasoned that trends in the shipping industry made a deeper harbor necessary. Although usually receptive to the GPA's development plans, the community responded with opposition to the news that the harbor should be deepened by eight additional feet. The GPA attempted to address these concerns by creating the SEG in January 1999 to include representatives from government, business, and the public directly and equally in a deliberative process of identifying the project's environmental impacts and formulating a plan to alleviate them. However, practice did not reflect this ideal.

In this chapter, I discuss the impact of the consensus-based stakeholder model on early SEG interactions through a rhetorical exploration of the second meeting held on February 2, 1999. I use the notion of public vocabulary to examine the characterizations, myths and key terms provided for the group by the consensus-based vocabulary and members' subsequent attempts to redefine these terms. Condit and Lucaites (1993) identify the public vocabulary as "the shared rhetorical culture" (p. xiv) that rhetors employ as they attempt to direct the community in its struggle to "negotiate its common needs and interests" (p. xii). I argue that the assumptions of equality, free and open deliberation toward the common good, and rational, authoritative consensus found in the consensus model produced a vocabulary that was inconsistent with actual group practice. Participants' desires to craft characterizations, myths and key terms that reflected group practice better and were more suitable for their task produced long debates and frustration, which constrained them from progressing in their mission. I begin with a discussion of the theoretical foundation of the consensus model and the practical workings of the model. I next overview the GPA's Harbor Deepening Project. I then discuss the notion of the public vocabulary and explore the workings of the consensus-based vocabulary in the SEG. Finally, I summarize the findings and discuss the implications of this study for public participation research.

THEORETICAL FOUNDATION OF THE CONSENSUS MODEL: DELIBERATIVE DEMOCRACY

Democratic theorists contend that the issue of deep disagreement has long been one of the challenges of the American political experiment (Bohman, 1997; Gutmann & Thompson, 1996; Hardin, 1999). In *Democracy and Disagreement* (1996), Amy Gutmann and Dennis Thompson attempt to offer a comprehensive solution to this problem. Gutmann and Thompson argue that in cases of moral disagreement between citizens and their representatives, opponents "should continue to reason together to reach morally acceptable decisions" (p. 1). Arguing that American society is suffering "from a deliberative deficit not only in our democratic politics but also in our democratic theory" (Shapiro, 1999, p. 32), Gutmann and Thompson outline a theory of deliberative democracy to guide this approach to disagreement.

Deliberative democratic theory begins with the assumption that "legitimate decisions require equality" (Bohmann, 1997, p. 321). Equality makes "the deliberative playing field . . . nearly level" (Gutmann & Thompson, 1996, p. 133). Although "differences in opinions, tastes, preferences . . . as well as in some resources such as knowledge" can exist, participants cannot possess "differences which make for disproportionate political advantage and persistent political disadvantage, such as differences in social circumstances and in basic public skills and abilities" (Bohman, 1997, p. 326). In this way, "deliberative arrangements work to dampen, and optimally to eliminate entirely any arbitrary inequalities" (Knight & Johnson, 1997, p. 288) among participants. Because deliberation can "diminish the discriminatory effects of race, class and gender inequalities" (Gutmann & Thompson, 1996, p. 133), it is superior to negotiation and bargaining.

Equal participants produce a deliberative process that is "free" (Cohen, 1997, p. 74) and "open to all" (Michelman, 1997, p. 160). Arguments are treated neutrally, unbiased "for or against any particular alternative" (Knight & Johnson, 1997, p. 292). The rational force of the better argument guides deliberation, ensuring the equality of all participants because "the use of material advantage in the political process" (Knight & Johnson, 1997, p. 295) is unnecessary. As deliberators offer their perspective, propose ideas and make compromises, they forge a "collective will, a joint intention" (Richardson, 1997, p. 360). This common good is "not identical with the wills of any one person" (Richardson, 1997, p. 358), but rather a notion crafted through debate and compromise.

Through this process "deliberators arrive at a rationally motivated consensus" (Cohen, 1997, p. 80). Because decisions are guided by rationality, "relations of power and subordination are neutralized" (Cohen, 1997, p. 78). Such decisions have more permanence and authority than majority rule because the "decisions of a majority at any particular time are provisional, since it may always be revisited by subsequent majorities" (Gutmann & Thompson, 1996, p. 28).

The apparent ease and straightforwardness of this theory has made it "the single most popular move of democratic theorists in our time" (Hardin, 1999, p. 103). Yet, critics have long found the core assumptions of equality, free and open deliberation to the common good and rational, authoritative consensus problematic.

Craig Calhoun (1992) concludes that the main problem with deliberative theories is that they ignore "the power relations, the networks of communication, the topography of issues and the structures of influence in the public sphere" (p. 35). According to Hardin, this notion of equality makes deliberation possible "only in the parlor room discourses or in the small salons of academic conferences, not in the normal world of rough and tumble politics" (1999, p. 112). Nancy Fraser argues that in addition to being unrealistic, the deliberative notion of equality actually masks "subtle forms of control" (1992, p. 119) and works "to the advantage of dominant groups in society and to the disadvantage of subordinates" (p. 120) because elite control of vocabulary and meaning excludes the less powerful from discussion.

Other critics question why "political debate ought to be focused on the common good" (Cohen, 1997, p. 72). According to Shapiro, rather than uniting individuals, deliberation brings "differences to the surface, widening political divisions" (1999, p. 32). G. Thomas Goodnight and David B. Hingstman (1997) argue that this is especially true in a society that is "noninstitutionalized, fragmented" (p. 352). Thomas McCarthy (1989) agrees that because the public is "socially, culturally, and psychologically diverse" (p. 128), the notion of the common good is unworkable. Because the common good is an impractical notion, critics argue that promoting it might actually facilitate exclusive deliberative practices (Koivisto & Valiverronen, 1996).

The diversity and fragmentation of society produces a situation in which there is "no common measure by which to assess the relative weights of reason articulated in different evaluative languages" (McCarthy, 1992, p. 65). With no common standard to judge alternatives, the notion of consensus becomes problematic. Patricia Roberts (1996) argues that the promotion of consensus causes the "oppression of steadfast dissenters" (p. 62). Mouffe (1999) similarly suggests the notion "that political questions can be decided rationally" (p. 753) can lead to the domination by the perspective with the most power.

In short, critics believe that "the search for a completely abstract definition of the public sphere could be part of the problem" (Hohendahl, 1992, p. 102). McCarthy suggests that we should "reconsider whether there might be a reasonable, nonviolent alternative to discursive agreement, hermeneutic consensus, and negotiated compromise" and develop "structures and processes of democratic public life . . . [that] reflect an awareness that unresolveable yet reasonable disagreements are also possible in principle" (1989, p. 152). Despite these criticisms, proponents of deliberative democracy continue to defend it as a practical solution to democratic disagreement (Gunderson, 1995; Gutmann & Thompson, 1996).

Researchers and practitioners in environmental participation have also begun promoting a similar approach. According to Crowfoot and Wondolleck (1990), these efforts differ from the traditional public hearings and comment periods because they stress "collaboration among contending interest groups instead of adversarial relationships [and] consensus decision making rather than judgments by authorities" (p. 1). They are united by "the structure of the process itself" constituted by "who is involved, how they are involved, and how issues are framed and then acted upon in making and then implementing decisions" (Crowfoot & Wondelleck, 1990, p. 22). These definitions of who, how and to what end are firmly grounded in deliberative democratic assumptions of public argument. Participants are described in the literature as stakeholders. Stakeholders are characterized as active because they are "willing and able to play a much more active role in environmental issues" (U.S. Environmental Protection Agency [US EPA], 1997a, p. 2) than average citizens. Stakeholders also "represent . . . a broad spectrum of interests" (Hendee et al., 1976, p. 134) including those interests previously underrepresented in environmental decision making. Finally, stakeholders are characterized as equal "partners . . . in the overall process" (Spyke, 1999, p. 17). These active, representative and equal participants are to enter into a "collaborative framework" (US EPA, 1997b, p. 3) and engage in a "community dialogue" (Farrell, 1999, p. 4) with each other. Through deliberation in this open forum, stakeholders "find there is more common ground than expected" (John & Mlay, 1999, p. 368). In this sense, the deliberative process "changes the scope and nature of the concerns that enter into decision making, as well as the weights those concerns are given" (Spyke, 1999, p. 20). Such discussions produce a sense of the common good or the "general public opinion" (Hendee et al., 1976, p. 134) of a particular community. This common ground "provides a basis for compromise of views and a workable consensus" (Committee on the Impact of Maritime Services on Local Populations [CIMSLP], 1979, p. 53–54). Because "all involved agree that everyone's concerns have been heard" (Spyke, 1999, p. 12), consensus decisions are inclusive, representative and more rational than decisions led by experts or by majority. According to Farrell (1999), consensus decisions "promise superior environmental performance and economic results" (p. 1).

In summary, the consensus model forwards a characterization of stakeholders as active, representative and equal, a narrative of group process in which free and open deliberation leads to a sense of the common good and a guiding goal of rational, authoritative, consensus decision making. However, as this abstract notion has been translated into practice, a number of issues have arisen that reflect the problematic assumptions of deliberative democracy. These issues include first deciding who speaks for whom and which interests are most important. Fundamental differences in stakeholder expectations and perspectives also create disagreements, which tend to make consensus decision making difficult. In turn, stakeholder efforts often become lengthy and expensive and result in the adoption of lowest common denominator solutions.

Despite suggestions that this model may be unworkable, in 1995 President Clinton identified it as the preferred participation mechanism at the federal level in his Reinventing Government Initiative. Reasoning that "better decisions result from a collaborative process with people working together" (Clinton & Gore, 1995, p. 1), Clinton supported consensus-based efforts in federal agencies such as the Environmental Protection Agency's [EPA] Project XL and Good Neighbor Dialogues and similar activities undertaken by the Department of Energy [DOE]. Because it promised improved participation through "inclusiveness and flexibility" (Spyke, 1999, p. 2), the consensus-based stakeholder model began to appear in the environmental efforts of industry and state and local government. It was also the participation mechanism adopted by the GPA in 1999 to deal with opposition to its proposed deepening project.

THE SEG AND THE SAVANNAH HARBOR IMPROVEMENT PROJECT

Ever since General James Oglethorpe settled the colony of Savannah in 1733, the port has been central to the livelihood and the economic prosperity of the city (Bell, 1977). Described as more of a "shallow river" than a deep "natural harbor" (Southern Environmental Law Center [SELC], 2000, p. 1), port authorities have engaged in a number of activities to deepen, widen and alter the river that created this harbor. These efforts, which began as early as 1852, consisted of closing and opening smaller channels, removing grassy knolls, constructing dams, turning basins, and installing a tide gate to control the infiltration of sand. The most significant changes, however, were four major deepenings that gradually took the 13-foot river to 42 feet by 1994 (Granger, 1968). To facilitate these efforts, the State of Georgia and the Army Corps of Engineers joined together to form the Georgia Ports Authority in 1948 (Sieg, 1985). Over the years, citizens generally supported the GPA's drive to craft a suitable harbor and often equated the increasingly polluted water with "*the smell of money*" (Russell & Hines, 1992, p. 175). Past support made the intense opposition brought about by the 1999 harbor-deepening project very surprising. In the following section, I trace the development of the opposition voiced by all sections of Savannah's society that culminated with the creation of the SEG.

Discussions of the 1999 harbor-deepening project began in 1997. The GPA argued that changes in the shipping industry made it necessary to deepen the harbor (GPA, 1998a). The agency reasoned that shipping lines were increasingly using large container ships rather than bulk or breakbulk transport. Containers were metal boxes that could be lifted from the ships and placed directly on the backs of trucks or railcars. Because container ships were larger than conventional ships, the GPA argued that they were constrained in the Savannah harbor. For example, "in 1996, more than 52 percent of container ships serving the Port of Savannah were operationally constrained due to inadequate channel depth" (Krueger, 1998a, p. 5). Problems included maneuverability and time lost waiting for the appropriate tidal stage (GPA, 1998a, p. 38). The GPA argued that ships were continuing to get bigger. In 1998, fifteen of the super ships "135 feet wide and 1,043 feet long—longer than the Eiffel Tower is tall . . . [with] a draft of 47.5 feet but requir[ing] . . . drafts up to 52 feet" (Seabrook, 1998, p. 1B) were in service. Given this situation, officials stated that "increased channel depths [were] necessary to accommodate the increasing drafts of these vessels" (GPA, 1998a, p. 9).

Larger ships also meant the increasing consolidation of trade routes. The GPA claimed that industry analysts predicted that consolidation of shipping routes would lead to the emergence of three megahub ports on the eastern seaboard with all other ports becoming secondary. With the northeastern ports in New York and Virginia securing the first two positions, Savannah was "locked in a battle with Charleston, S. C. to be named the third" (Schmitt & Guidera, 1998, p. 1). With Charleston engaged in a deepening effort, port officials argued that if Savannah did not deepen it might be relegated "to 'feeder' port status" (Editorial: Project should proceed, 1998, p. 1).

The GPA argued that due to these two trends, the harbor must be deepened for the economic prosperity experienced in recent years to continue (Conflict history, 1998). In 1997, the port handled a record total of cargo and "directly or indirectly support[ed] 80,100 jobs, [was] responsible for $1.8 billion in wages, generate[d] $23 billion in revenue and account[ed] for $585 million in state and local taxes annually" (Editorial: Project should proceed, 1998, p. 1).

Based on these justifications, the GPA took the opportunity to perform a project feasibility study granted by Section 203 of the Water Resources Development Act [WRDA] of 1986 in March 1997 (GPA, 1998a, p. 9). This section enabled the community to petition the Corps regarding a perceived water resource-related need and then to undertake the feasibility study work. The Corps would act as an advisor, help to secure authorization, and split the cost of the project. The GPA released its five-million-dollar Tier I Draft Feasibility Study [FS] and Draft Tier I Environmental Impact Statement [EIS] in May 1998. In the two "half-foot thick"

(Krueger, 1998a, p. 2) documents, the GPA concluded that deepening the harbor to 50 feet was environmentally feasible and would produce $35 million in annual net benefits.

Despite the previously popular emphasis on potential economic benefits, the release of these documents drew responses from nearly every impacted group (Krueger, 1998b). An extended 30-day comment period yielded comments reflecting economic and environmental concerns.

The Southern Environmental Law Center (SELC) led opponents such as the Coastal Environmental Organization (CEO) and the Nature Conservancy in arguing that harbor deepening was not necessary to maintain harbor economic viability (Editorial: Politics and water, 1998). These groups were amazed that the agency would pursue another deepening only four years after a similar project. They asserted that the GPA's rush to get the project approved in the WRDA of 1998 reflected a "fast track" mentality that resulted in a flawed economic analysis (Seabrook, 1998, p. 1B). They claimed that the GPA overestimated the need for deep draft harbors and failed to consider the interplay between the Savannah Port and others (Taxpayers For Common Sense, 1999).

Resource agencies such as the Fish and Wildlife Service (FWS) and the EPA joined with local environmental groups to express a number of environmental concerns. First, these groups claimed that the drastic mitigation measures outlined in the study would facilitate "increased salinity and decreased dissolved oxygen in the river, causing catastrophic collapse of the striped bass fishery, and the habitat for the endangered shortnosed sturgeon, while also causing the loss of over half the tidal freshwater marsh which forms the centerpiece of the Savannah National Wildlife Refuge" (SELC, 1998). The striped bass, the shortnosed sturgeon and the Wildlife Refuge had all experienced tremendous impacts over the past few decades due to high pollution levels and river engineering efforts. For example, as a result of the tide gate, the Refuge lost approximately 2,800 acres of freshwater marsh and both fish populations had been reduced by half (Krueger 1998a; Seabrook, 1998). Since all three had begun to recover, concern that deepening would create a similar scenario was great. FWS biologist John Robinette concluded that deepening would produce "a disaster that cannot be reversed. It's kind of scary" (Krueger 1998a, p. 1).

The second environmental concern was the historically sensitive issue of water quality. In the early 1970s, Savannah had been criticized by environmental advocate Ralph Nader for its practices of dumping untreated sewage in the river and allowing local industry to dispose of waste similarly

(Fancher, 1976). Georgia was also experiencing a drought and engaged in "water wars" with Alabama and Florida over water rights (Editorial: Politics and water, 1998). City officials were troubled that increased salinity caused by deepening would limit their ability to provide the citizens with fresh water. Local industry also worried that this might increase water treatment costs (Krueger, 1998a). Beneath the river was also a freshwater aquifer that was experiencing saltwater intrusion as a result of unlimited withdrawal in the past. Deepening might have punctuated the protective layer and caused further saltwater contamination (Cochran, 1998).

The final environmental concern was sand erosion at Tybee Island. The home of Savannah's only beach, the small island was shrinking as a result of past harbor projects (Kreuzwieser, 1995, p. 30). Residents were concerned that further deepening would increase erosion. The GPA maintained that "there'll be plenty of time to address concerns about the project" (Sechler, 1998, p. 1).

In light of these reactions, the GPA decided to temper the proposed dredging depth from 50 to 48 feet and created the SEG as a mechanism to ensure "two-way communications" (GPA, 1998b, p. 1) between the GPA and interested parties throughout the project process. Specifically, the group was to represent organizational interests in developing "an environmentally acceptable mitigation plan" (GPA, 1998b, p. 25) for the project. Membership was open to "government, public and private" organizations (GPA, 2000, p. 4). However, the GPA did specifically identify groups such as the EPA, FWS and Georgia Environmental Protection Division [EPD] as necessary participants. Although 1998 passed without Congressional authorization, the GPA continued to move forward with the project and convened its first SEG meeting in January 1999.

PUBLIC VOCABULARY

Because this chapter seeks to critique the ways in which the consensus-based model impacts public argument in environmental participation, a methodology is needed that allows me to explore both the underlying assumptions of the SEG (ground) as well as their impact on subsequent discourse (figure). Condit and Lucaites' (1993) public vocabulary concept is a suitable tool for this study because it enables the critic to consider "the efforts of individual rhetors in specific groups and organizations [figure] within the rhetorically constituted matrix of ideological commitments [ground]" (p. 16). In this section, I define and discuss this concept.

According to McGee, a "vocabulary of complex, high-order abstractions that refers to and invokes a sense of 'the people'" (1980/1995, p. 452) defines a particular collectivity. This vocabulary consists of simply "a group of *words*" (p. 447) entitled ideographs. An ideograph is an "ordinary-language term found in political discourse . . . a higher-order abstraction representing collective commitment . . . to a normative goal" (p. 452) such as liberty or equality.

Condit and Lucaites (1993) expand this notion of a public vocabulary, suggesting its use as "a general mode of the rhetorical process of public argumentation" (p. 16). Condit and Lucaites identify the public vocabulary as "the shared rhetorical culture" (p. xiv) that rhetors employ as they attempt to direct the community in its struggle to "negotiate its common needs and interests" (p. xii). Although this vocabulary includes the "full complement of commonly used allusions, aphorisms, characterizations, ideographs, metaphors, myths, narratives and topoi or common argumentative forms" (p. xii), these scholars agree with McGee that ideographs are "the central, organizing elements for any rhetorical culture" (p. xiii). Condit and Lucaites note that, in addition to "represent[ing] in condensed form the normative, collective commitments of the members of the public . . . [ideographs] typically appear in public argumentation as the necessary motivations or justifications" (p. xiii). In this sense, they act as "the moral of the story in public political narratives" (Condit, 1987b, p. 3).

Public political narratives are "story forms" that recount "a thing being done, or supposed to have been done, which is adapted to persuade" (Condit, 1987b, p. 4). When they gain "persuasiveness and force among an active plurality of persons in an active discourse group" (p. 4) they are transformed into public political myths. Myths, in turn, derive their meaning from characterizations. Characterizations are "universalized descriptions of particular agents, acts, scenes, purposes or agencies which, when they become culturally accepted as accurate descriptions of a class can be labeled character-types" (p. 4). Examples of character types in American culture are "'Boy Scouts,' 'politicians,' and 'the South'" (p. 4).

Through public argument, ideographs, myths, characterizations and other elements of a public vocabulary interact to develop a set of relationships that provides "a rhetorical foundation of public value" (Condit, 1987b, p. 3) for a particular community. Such foundations act as "substantive constraints on precipitous social change. At the same time, the particularized character of their formation presumes that the public vocabulary must receive at least partial revision when new conditions arise" (Condit & Lucaites, 1993, p. 3). In this sense, the elements of a rhetorical culture maintain somewhat of a discursive constant, providing lenses through which a critic can "gain a reliable, if indirect, indication of the patterns of interest and forces supporting and constituting the public discourse" (Condit, 1987b, p. 2).

CHARACTERIZATION, MYTH, AND IDEOGRAPH IN THE SEG'S EARLY DISCOURSE

The consensus-based model crafts a vocabulary in which stakeholders are characterized as active, representative and equal; group process follows a myth of free and open deliberation to the common good; and consensus acts as an ideograph meaning rational, authoritative agreement. According to Condit (1987a), "once universal terms, narratives, and characterizations are created and supported, they carry a force separate from the wishes of the collectivity" (p. 87). Once articulated by researchers and practitioners, the vocabulary of the consensus-based stakeholder model takes on its own rhetorical force.

When the GPA adopted the consensus framework, its vocabulary served as a forceful rhetorical foundation for group discourse. In the opening of the second group meeting, facilitator Morgan Rees drew on this foundation to describe the SEG process. Rees began by characterizing all stakeholders as equal with "everybody" having the "chance to speak up and be heard" (Stakeholder Evaluation Group [SEG], 1999, side 1). This characterization acted as a foundation for a "story" (Condit, 1987b, p.4) of a free and open deliberative process where "everyone ha[d] a chance to say what their concerns are" (SEG, 1999, side 1). With "everything on the table" (side 1), the group could debate to determine the common good. However, because this language was inadequate to capture the reality of group interaction, ensuing articulations of consensus vocabulary in this meeting produced attempts to "question or disarticulate" (DeLuca, 1999, p. 41) its terms. These questions created rearticulations of the consensus vocabulary. In this way, the vocabulary placed "substantive constraints on precipitous social change" (Condit & Lucaites, 1993, p. 3) in the early discourse of the SEG. In this section, I explore this rhetorical struggle in the SEG's discussions regarding their position in the GPA deepening process and the definition of a stakeholder in the second meeting held February 2, 1999.

The Position of the SEG in the Deepening Process

Discussion over the SEG's position in the deepening process first arose during a debate over the January meeting minutes. In January, the GPA had identified the endangered shortnosed sturgeon, striped bass, salinity, dissolved oxygen, and chloride distribution as the five principal topics suitable for the SEG. Stakeholders had indicated that there were other topics identified in the public comments and that these comments were the very basis for the SEG's existence. Members suggested that debate should begin with these issues. Although GPA representatives asserted that these issues had been adequately addressed, stakeholders demanded that the GPA produce a document cataloging all comments and how the GPA had addressed them. The GPA offhandedly agreed. In the February meeting, this issue arose again. This discussion revealed inconsistencies between group practice and the consensus vocabulary articulated by Rees at the beginning of the meeting.

One unidentified member asked the GPA whether or not they were going to provide the requested catalogue. Because performing the myth of the common good began with putting all issues on this table, this list was essential. Based on this information, the SEG members could determine which "issues should be included in this effort by the SEG" (SEG, 1999, side 1) and define the common good. Responses from the GPA's representative revealed an inconsistent narrative. First, the GPA consultant Ed Modzelewski argued that the document was not necessary. He asserted that the public comments contained in the Tier I EIS were "historical issues" that would require "going back and looking at the comments on the Tier I EIS" (side 1). By framing certain topics as historical issues, this GPA representative worked to set the agenda for the SEG rather than allowing members to begin with a list of all issues. Project director Larry Keegan then refused this request by asserting that "it would take an awful lot of time for us to go through and cover what is in appendix H of the final EIS. We are not prepared to do that today" (side 1). By refusing to prepare this document, the GPA effectively limited the issues to be addressed by the SEG. Control by one group contradicted the myth of the common good where the good was determined through free and open debate.

Facilitator Rees worked to mend the inconsistency between the GPA's actions and the common-good myth. He first reasserted the myth by reasoning that since all issues would be "put on the table . . . all of those issues will get adequately addressed" (side 1). Rees also argued that since "the objective of this group is to make sure that all the issues are adequately addressed . . . production of a large document or a lengthy briefing at this time by GPA may be premature" (side 1). Rees attempted to assure participants that the GPA's refusal was not an attempt to control their agenda. Because deliberation was free and open, all issues would be addressed eventually. Rees' rearticulation of the myth drew immediate support from another member who agreed that "the function of this group is to identify all possible issues and then figure what to do about them, rather than have GPA explain why they did this or that" (side 1). Here, the myth of the common good legitimized the GPA's attempt to limit the group's agenda and control deliberation. The GPA could control the agenda because as the group participated in the myth of the common good, all issues would be addressed.

This attempt to smooth over inconsistency with the myth of the common good did not gain the assent of all group members. Some stakeholders responded by forging a new narrative of group practices. Sam Drake of the FWS argued that, according to his "recollection" of "some of the 101 meetings," the outline of group activities was "a little different" (side 1). Drake recounted a narrative in which the GPA controlled the SEG's agenda because they were "gonna seek congressional authorization" and wanted to "not complicate the process" (side 1). By offering an alternative narrative of group deliberation in which the GPA defined the topics of debate, Drake disarticulated the myth of the common good. Stakeholder Tom Meronek from the Georgia Department of Natural Resources continued the tale by narrating the story of the last meeting when the entire SEG "created a list of key issues that we felt we should address early on" (side 1). Recognizing the GPA's control of the agenda, he noted that these issues were not listed "in the minutes at all" (side 1).

Following stakeholder attempts to describe a realistic narrative of group process in which the GPA controlled the common good by limiting the SEG's agenda, Rees again worked to fix the consensus-based myth. He argued that because stakeholders represented all issues, "this group is going to deal with them in one way or another" (side 1). Reasoning that the SEG would set its own agenda and determine the good through a debate of these items, Rees let the GPA "off the hook" (side 1) for producing the document. Paradoxically, the myth of the common good legitimized the GPA's control.

Discussion of the SEG agenda emerged again as the GPA project director Charles Griffen presented an overview of the Tier II process. Griffen outlined three phases: the detailed design phase, which contained an economic analysis, some environmental analyses, and engineering studies; the environmental evaluation phase, which was the domain of the SEG; and the cultural resource phase, which handled Ft. Jackson, the old military fort, and the USS Constitution, a Civil War ship sunk at the bottom of the harbor.

As Griffen discussed the three phases, he clearly distinguished between them, limiting the SEG's agenda and revealing inconsistencies between group practice and myth. Griffen limited the SEG's agenda by arguing that the studies in the detailed design phase were not "deemed . . . to be facilitated or necessary to be facilitated" (SEG, 1999, side 2). Griffen asserted that although some of these items were specific concerns of the SEG members such as the Tybee Beach evaluation and the economic analysis, they were part of the "unfacilitated process however . . . [with] a very strong public input component" (side 2). Griffen removed these items from the SEG agenda by categorizing them as part of the unfacilitated phase. He next outlined "the fundamental elements that are in the domain of the facilitation [or SEG] process" (side 2). On this agenda were "four or five elements" such as "dissolved oxygen and salinity and chlorides" (side 2). This move confined the SEG discussion only to these elements. Because the myth of the common good depended on beginning with all issues, Griffen's attempt to limit the SEG agenda revealed an inconsistency between myth and practice. Griffen rationalized this inconsistency by arguing that it was necessary to produce "the most desirable design, balancing the environmental issues, balancing . . . the market, economics that go with it and balancing constructability" (side 2).

Recognizing Griffen's control of the SEG agenda, activist Ben Brewton questioned, "you're explaining parts of the project that are beyond the scope of the stakeholder's group?" (side 2). Sensing that their control of the agenda was revealed, the GPA representatives quickly rearticulated the myth of the common good. First Griffen stated, "I'm explaining the parts of the project in its total environment. I'm going to come back. I'm gonna go through this and then I'm gonna come back and show where the SEG, the things that this group has up to this time decided is a part of what they want to do" (side 2).

Griffen attempted to reestablish the myth of the common good by asserting that the SEG had actually determined the limited list of issues he outlined. Facilitator Rees reasserted "what was said in the formation of the group" as he jumped "in as facilitator . . . and said . . . any issue that the SEG wants to bring to the table is a legitimate issue for the SEG" (side 2).

By rearticulating an essential component of the common good myth—that deliberation began with recognition of all issues—Rees worked to reestablish this myth.

Brewton responded to this rearticulation with another attempt to craft a more realistic narrative of group interaction. He stated, "ok, well, we'll listen; it just appeared to me that many of these things [considered not appropriate for the SEG] are the very essence of what the stakeholders are" (side 2). In this comment, Brewton made the GPA's control of the SEG agenda explicit. Griffen instantly disregarded this dearticulation with the statement, "ok, I appreciate your comment. We are in that portion of the area that you saw during the public information meeting, detailed design . . ." (side 2). Griffen intimated that this description was inaccurate by closing down the discussion and moving on.

The second issue to emerge regarding the position of the SEG in the deepening process was the meaning of consensus. The previous struggle over the SEG agenda caused one member to question this meaning. Since consensus functioned as an ideograph, it provided the motivation for action in the myth of the common good. Questioning of the suitability of this myth for describing the SEG interaction led to questions regarding the meaning of consensus.

Mitch King asked, "You indicated that you're gonna do a NED; is that going to cover all the costs of an identified mitigation of proposed alternatives?" (side 2). A National Economic Development (NED) analysis was an economic study performed by the project proponent that identified which project alternative had the greatest benefit-to-cost ratio. According to federal guidelines, the deepening alternative resulting from this analysis should be the option chosen for the project. However, a community could choose a different alternative in the case of a compelling reason, such as if the cost of mitigating against extreme environmental damage made the selection of a lower benefit-to-cost ratio necessary. By asking whether the GPA would follow the SEG recommendation even if it produced lower benefits, this stakeholder questioned whether the SEG consensus had authoritative meaning or not. King followed an assurance from Keegan that "it has to" with a direct question to determine the authority of the SEG consensus. He questioned,

> [Y]ou felt like Georgia Ports Authority wanted to accept the recommendation of this group, which may be different than the NED recommendation has that been agreed upon. That whatever this group

> identifies as the recommendation the Georgia Ports Authority's gonna go with that? (side 2)

As a product of the myth of the common good, consensus decisions represented rational and authoritative conclusions. By questioning whether the SEG consensus would have more authority than the NED in the GPA's actions, King attempted to determine if this meaning was being practiced.

Because consensus acted ideographically, its establishment as rational and authoritative was essential to the performance of the common good myth, so GPA representatives worked to fix this meaning. First, Rees tried to stop discussion by arguing that the question was "a new item and I don't know what the real answer is" (side 2). Griffen then began rearticulating the consensus vocabulary by characterizing stakeholders as free and equal because all had "a place at the table" (side 2). This attempt to misdirect King did little to appease him. King again questioned by asking if the "Georgia Ports Authority made a commitment and basically said whatever this group agrees upon as far as a depth and an alternative, we'll live with that even though it may not be the most economically viable alternative but it might have some other considerations such as environmental considerations" (side 2). By questioning whether the SEG consensus decisions would have authority over the GPA's actions, King attempted to determine whether practice was consistent with the ideographic meaning of consensus.

Griffen again attempted to table discussion by making this question appear irrelevant. However, environmental advocate Ben Brewton became frustrated and attempted to force Keegan to define consensus by asserting, "I think it's a good question, it goes to the very core of what some of us in this group are concerned about, and it's a question that should be able to be answered with a yes and a no" (side 2).

Here Brewton recognized that the definition of consensus was essential to group activities because a definition as rational and authoritative would yield a plot of the common good, whereas a different definition would produce a different narrative. For this reason he pushed Griffen to define the term.

Ports Authority Deputy Director David Schaller responded to this pressure by stating,

> The work of the SEG is important to the Georgia Ports Authority. I can confirm that. Uh, we are committed to the process and committed with the SEG and the results of the SEG in terms of the final

> recommendation will be taken under consideration to the very fullest extent when the time comes. (side 2)

Terms such as "committed" and "taken under consideration to the fullest extent possible" intimated that consensus had authoritative meaning. Yet the response also suggested that in practice the GPA could disregard the SEG consensus. Recognizing this inconsistency, Brewton responded,

> David, thank you for the very statesman-like and very carefully word-smithed answer . . . We appreciate that, but I think that the problem that we have here is that we keep getting mixed signals. Mr. Keegan just said something a few minutes ago . . . that's very clear; he said GPA would follow the recommendations of this group. What you are saying now is that GPA is going to consider the recommendations of the group, and there's a big difference there. (side 2)

Brewton's response reflected recognition of inconsistency in the meaning of consensus. At one point the GPA defined this term as authoritative and in the next minute the term was considered to be merely a recommendation. In the face of this inconsistency, Brewton continued to push for a meaning of consensus that accurately reflected its practice. He stated,

> I think the people here, the stakeholders, are here in good faith; we want to participate, we want to work with you, but we don't want to have the rules changing as we go and we really want to hear from the GPA not a carefully wordsmithed answer but an honest from the gut, from the heart answer. (side 2)

Brewton's call for a consistent definition of consensus revealed a desire for a vocabulary more reflective of actual practice.

This argument propelled Schaller into providing a realistic definition of this term. According to Schaller, the SEG consensus could not have complete authority over GPA actions because "the question of what the final depth chosen by the Georgia Ports Authority is, is a matter of policy that will be considered by the Georgia Ports Authority board of directors" (side 2). Schaller admitted that SEG consensus meant merely a recommendation rather than an authoritative decision.

Grateful for a realistic definition of consensus, Brewton responded, "Well, I think that clarifies in a good enough way that you are going to consider this group but not necessarily be bound by it. And that's fine. I just don't think you should be saying one thing in one occasion and not in another" (side 2). Likewise, King "appreciate[d]" the effort to "make it clear" (side 2). In this interchange, the emergence of a meaning of consensus, which accurately reflected group practice rather than the continued rearticulation of the consensus-based vocabulary, satisfied participants.

The Definition of Stakeholder

Debate regarding the definition of *stakeholder* emerged as the Modeling Technical Review Group (MTRG), a technical advisory committee created by the GPA prior to the first SEG meeting, that updated the group on its activities. The GPA reasoned that the immediate creation of this committee, which was responsible for the design and implementation of scientific studies to be used to create a model for the prediction of chlorides, dissolved oxygen, and their interaction at various depths, was crucial because these studies demanded springtime conditions. Committee membership was a mixture of engineering experts from the GPA's consulting firm, Applied Technology and Management (ATM). Discussion of the MTRG's activities revealed that this subgroup was distinct from other stakeholders and conducted closed deliberations.

In his update, ATM president Ed Modzelewski crafted a characterization of stakeholder as equal and representative. Modzelewski reported that participants at the first MTRG were "representative [of] different agencies, the Corps . . . John Sawyer from the city . . . Atlanta, [and] EPA" (side 2). In this sense, the participants of the MTRG were representative of the SEG. During the meeting, there was "a lot of interaction" and "all of the group" (side 2) participated in the myth of the common good to reach consensus on a study design with input from "almost everyone in the group" (side 2). Modezelewski then questioned if there were "any details you want to bring up" (side 2).

Sensing that this articulation was inconsistent with practice, Savannah Manufacturer's Council representative Bob Scanlon questioned "where" (side 2) the group met. When he found that the group met in Atlanta, Scanlon suggested that this might reduce its representation because "there's been no state participation at all for either Georgia or South Carolina" (side 2) and the offices of both state agencies were near Savannah. Scanlon's suggestion initiated a lengthy discussion replete with reassertions that the activities of the MTRG ensured the equality and representation of all stakeholders. Members of the MTRG asserted that they were committed to "outreach" (side 2) in the form of "Web page" postings, "hard copies" "written comments" and e-mail and telephone interactions (side 2). Such activities ensured that all interests were represented. Because all members had access by participating "electronically if they couldn't make it to the meeting" (side 2), all stakeholders acted equally. The result was the "significant involvement" (side 2) of all parties in designing the studies (common good).

Throughout deliberation, it became clear that the members of the MTRG enjoyed a special status among stakeholders. First, discussion revealed that some members of the group were considered more important than others were. To justify the Atlanta meeting location, Ed Modzelewski argued that

> [T]he major thrust of the first three tasks are water quality issues and the keeper of those standards are really housed in Atlanta—its EPA and EPD. We wanted to accommodate that group in Atlanta primarily because they have busy schedules and they're really the two important agencies that really have to get involved. (side 2)

As debate continued, the preferred status of these agencies surfaced. The EPA, EPD and FWS were continually characterized as "primary agencies" (side 2) with "significant interests" (side 2). In contrast to other stakeholders, their participation was described as "at least very important and perhaps even critical" (side 2). This characterization of the EPA, EPD, and FWS revealed that in practice stakeholders were not all equal.

Second, discussion indicated a distinction between the MTRG committee as a whole and other stakeholders. To justify the Atlanta meeting location, Project Manager Charles Griffen classified the MTRG as a "scientific group" of "scientific folks" who "as with any engineer . . . don't necessarily need meetings" and could "adapt their activities as a scientific group to interact together in the best way that that group can deal with" (side 2). Griffen's description of the MTRG as a specialized group was inconsistent with the characterization of stakeholders as equal. As he continued, he also suggested that this group controlled SEG decision making because they would "bring back to the SEG the results of their deliberations and ask for—ask for SEG consensus about what they brought back" (side 2). In this narrative, as the technical elite determined the common good, other members of the SEG were reduced to the role of saying, "hey, . . . what they've got is the right thing" (side 2). This process was inconsistent with the common good myth.

As inconsistencies between stakeholder characterization and narrative and actual practice appeared, members began to demand a vocabulary more reflective of group interaction. Activist Ben Brewton identified inconsistencies between the guiding myth of the common good and actual group practice by recounting how he was excluded from the MTRG meeting. Brewton began to interrogate the narrative of group activities by explaining that he sent a representative to the MTRG meeting because he could not attend. He also asked the GPA representatives to "be copied on meetings correspondence, attendance lists and so forth" (SEG, 1999, side 3). The response he received, "that neither we nor the representative we asked to go were officially considered members of that group and as such we would not be copied on anything" (side 3), indicated that participation in the MTRG deliberations was not open as the myth suggested. Seeking a narrative that more clearly described group interaction, Brewton asked, "Can someone address if that information will be made available generally to those of us who request it or not?" (side 3). When this request was met with the GPA's representative Charles Griffen shuffling through his papers for his e-mail response, Brewton continued to push for an answer "in plain English rather than reading the message" (side 3).

Griffen gave a prepared response that suggested that although the MTRG was separate from the SEG in that they were "technical and scientific modeling professionals," the SEG had access to their deliberation and could comment on their decisions. Brewton highlighted the contradictory nature of this response with the statement, "I don't understand if he answered yes or no—we'll be made available copies of minutes, attendance sheets and findings or if we did . . . I just don't get the answer" (side 3). However, his attempt to force a dearticulation of the common good myth only initiated its rearticulation by the GPA Deputy Director David Schaller. Schaller asserted "there isn't any data we choose to hide from anyone" (side 3) and that any data was available "to everyone who wants the data" (side 3). Because information was available to all, deliberation was free and open. He then argued that information dispersal was not up to the GPA but rather to "the consensus of the working group" (side 3). In this sense, MTRG activities were consistent with the consensus-based myth and ideographic meaning. In an attempt to fix this rearticulation, Griffen suggested "we table this discussion" (side 3).

Following this interchange, Brewton pushed the group to discuss directly the definition of stakeholder. Brewton began his interrogation of the stakeholder characterization with a description of how he "specifically sent an e-mail in advance of that [MTRG] meeting asking that . . . a consultant that we had contracted with be included, uh when he attended the meeting he was told that neither of us were considered to be on the list or part of the committee" (side 3).

By highlighting that neither he nor his representative was allowed membership in the MTRG, Brewton suggested that in practice there was a distinction between stakeholders. He then questioned the consistency of this practice with the group vocabulary by asking, "I'd like to know why we seemingly have an open invitation for participants, yet when we asked that, we were told that no we weren't participants or members of the committee?" (side 3).

Brewton's recognition of this inconsistency initiated a series of attempts to dearticulate the consensus-based characterization of stakeholder. First, the GPA representative Charles Griffen began to recognize explicitly inequalities between stakeholders as he asserted that there was a "difference between the deliberations of the SEG and the deliberations of the technical task people" (side 3). However, he quickly caught his slippage from the standard stakeholder description and responded that this was "not the forum in which I care to deal with it" (side 3). Next, Tybee Island representative Bill Farmer crafted a characterization of stakeholder that more accurately captured diversity and difference between participants. According to Farmer, rather than all stakeholders being uniform,

> [S]ome are very interested in the water quality work that's being done by the technical review group, some are not, some are interested in other items and will not plug into that and will not care too much about that information. Others will care a lot about it and want to review it and take a long time to review it and understand it. (side 3)

Farmer's characterization identified differences among stakeholder interest, representation and power at various points in the deliberation process. This characterization was much more reflective of actual group experience.

In response to these alternative portrayals of the stakeholder, facilitator Rees quickly made an attempt to reassert the notion that all stakeholders were equal. He began by rearticulating the myth of the common good. In a seemingly "off topic" statement he described the SEG forum as "open . . . public involvement government" (side 3) where "you can't close the door to anybody" (side 3). Because all were allowed to participate in deliberation, a stakeholder in this myth was "anybody . . . [who] wants to join" (side 3). In essence, all who wanted could represent their interest in this forum equally. In the end, Rees was able to move the SEG to this meaning by moving discussion to another topic.

CONCLUSION: IMPACTS OF THE CONSENSUS MODEL ON THE SEG

When the GPA realized the staunch opposition that existed regarding their 1999 deepening project, the agency chose to implement a consensus-based stakeholder group that had been identified by researchers as the next panacea for environmental participation (Rabe, 1991). With goals of collaboration, cooperation and participation, the model seemed the most appropriate approach for a group created to address the concerns of citizens, government, and industry by involving them in impact studies and mitigation plan development.

During the second meeting, held in February, representatives of the GPA and the facilitator worked to establish a group based on the assumptions of equality, representation, open deliberation, rationality, and agreement found in the consensus model. During this meeting, it became apparent that this cooperative effort was somewhat unrealistic. The GPA acted inconsistently by distinguishing between "primary agencies" (SEG, 1999, side 2), "scientific folks" (side 2), and other members, excluding less knowledgeable stakeholders from the technical deliberations of the MTRG and disregarding consensus decisions. Recognizing these inconsistencies, perplexed stakeholders worked to craft a vocabulary that better reflected practice. Such dearticulations produced a more realistic account of group activities in which stakeholders were different in terms of status and interest, deliberation and decision making was controlled by the GPA and technical elite, and consensus decisions were reduced to recommendations. In response to such challenges, the GPA and the facilitator simply rearticulated consensus-based vocabulary. In this way, this vocabulary was often used to cover up and smooth over recognition of inequality and control. But because it was inconsistent with group practice, its continued articulation in the months ahead only fueled stakeholder frustration and decreased group productivity.

By the summer, Congressional appropriation and ensuing project approval by the Army Corps of Engineers created pressure for the group to move forward. Members employed a variety of rhetorical strategies to challenge the inappropriate vocabulary and instill more workable meanings. Adopting this more efficient vocabulary in which technical and agency elite directed the study process not only made the SEG more efficient but also made it resemble a traditional environmental decision-making group.

In this way, the SEG ultimately transformed itself from an egalitarian consensus-based group into an exclusive and efficient decision-making group to fulfill the first half of its mission. Efficiency and expertise replaced cooperation and participation as criteria for decision making (Williams & Matheny, 1995). The use of this model for the SEG led to a paradox. To ensure participation, the SEG had to sacrifice efficiency but to become efficient they had to sacrifice participation.

IMPLICATIONS FOR PUBLIC PARTICIPATION

There is an increasing battery of research indicating that these grand communitarian-type models are doing little to better the state of American politics (Burgess & Burgess, 1995; Cohen, 1997; McCarthy, 1992; Mouffe, 1999). A number of researchers in the area of environmental participation now assert that this latest panacea is completely unworkable in its present form (Amy, 1987; Rabe, 1991; Williams & Matheny 1995). The failure of the consensus model in the SEG supports this conclusion. Specifically, critics are concerned that the model attempts to erase the intensely political context of these efforts (Clary & Hornney 1995; Glover, 2000; Stephenson & Pops, 1991). Amy contends that environmental participation "is not simply about communication. It's also about power struggles. It is not only about common interests, but also about conflicting interests. And it's not only about horse trading but competition between conflicting values and different moral visions" (Amy, 1987, p. 228). According to Stephenson and Pops (1991), participation efforts "will inevitably occur in intense political environments and will themselves emerge as another form of politics" (p. 24). For this reason, it is a "mistake" to view consensus approaches "as the solution" (Amy, 1987, p. 226) to the problems of participation.

Because the consensus model is clearly an unworkable solution that fails to account for the inevitably political nature of environmental public participation, the continued development of this "detached and abstract" (Kolb & Associates, 1994, p. xiii) model should be abandoned for a more suitable approach. Critics of the model in participation research point toward one possible avenue. They suggest a "shift away from the myths of mediation to an open acknowledgement of the problems of practice" (Kolb & Associates, 1994, p. 492). They suggest for researchers to begin focusing on actual practice, identifying and exploring the specific strategies that worked for participants in particular efforts (Amy, 1987). Given the prevalence of power imbalances in this context, research would pay careful attention to those strategies that empower "weak or powerless groups to negotiate their own interests and rights" (Landais-Stamp, 1991, p. 282). Geertz (1973) and Leff (1980) likewise contend that a shift in focus to the inspection and interpretation of specific events provides an "effective antidote" (Leff, 1980, p. 339) to the pursuit of vacant theories. In this approach, theoretical concepts get developed through continued application. Studies borrow theoretical ideas from other studies and these concepts are "further elaborated and go on being used" (Geertz, 1973, p. 27) if they lead to new ideas or "if they cease being useful . . . [they] stop being used and are more or less abandoned" (p. 27). Theory is not built in "the form of a ladder leading up and down from higher order abstractions" (Leff, 1980, p. 347). Instead, theoretical ideas move "across a discipline as a fire burns through a forest, growing, shifting and receding in irregular patterns, gathering intensity from the matter it consumed, but having no existence apart from that matter" (p. 348).

This approach to theory building would certainly help move participation research from the relentless pursuit of the best model to a deeper understanding of what "in this time or that place, specific people say, what they do, what is done to them" (Geertz, 1973, p. 19). Rhetorical critics could contribute to the project of a grounded theory of participation by uncovering strategies that negotiate and challenge power imbalances and contribute to collaboration in specific cases. Strategies identified in one study could be explored, extended, or abandoned in other studies. These findings could inform findings in other research areas to develop a grounded theory of environmental participation.

CHAPTER 10

"FREE TRADE" AND THE ECLIPSE OF CIVIL SOCIETY: BARRIERS TO TRANSPARENCY AND PUBLIC PARTICIPATION IN NAFTA AND THE FREE TRADE AREA OF THE AMERICAS

J. ROBERT COX

> Flawed as they now are, nation-states are the only institutions in the world currently capable of acting as a buffer between the world's citizens and the lawless anarchy of the global economy. However, the only way we are going to re-claim and democratize the state is with an active civil society engaged in real participatory democracy.
>
> —Maude Barlow, *The Future of Nation-States*

Canadian social critic Maude Barlow's call for civil society to "re-claim and democratize the state" (2001a, p. 4) reflects globalization's ironic reversal of a tenet of democratic theory. In its classic formulation, civil society and the state exist in tension, and the weakening of state power is "regarded as helpful to democratic consolidation," that is, in strengthening the freedoms of association, speech, and political mobilization (Kim, 2000, p. 123). A sphere of public expression arising from civil society[1] thus mediates state power and private interest, and is a prerequisite for democratic governance. In particular, civil rights, environmental protection, and labor policies have emerged often as the result of, and only after, vigorous debate among the contending forces within civil and political society.

With recent developments in global economic integration (precipitated by the so-called 1994 Uruguay Round of trade negotiations[2]), however, some critics argue that nonpublic administrative arrangements (such as private dispute panels) are filling the public spaces of democratic governance. These developments pose troubling questions for civil society, threatening to disrupt the "participatory, consultative, transparent, and publicly accountable" nature of democratic states (Scholte, 2000, p. 262). Multinational agreements such as the General Agreement on Tariffs and Trade (GATT) and its successor, the World Trade Organization (WTO), for example, have moved beyond traditional concerns of the border (tariffs and quotas) to the heartland of popular governance. Under expansive rules that protect the investments of foreign businesses, state actions that regulate worker safety, public health, or environmental quality—to the degree they affect investment income—may be determined to be *nontariff* barriers to trade and overturned or the state penalized with trade sanctions. For this reason, some critics argue that, despite the claim that "free global markets promote democracy,"[3] regional trade agreements, as well as the policies of the International Monetary Fund and World Bank, have led to a "democracy deficit" (Scholte, 2000, p. 267).

As a consequence, some critics of globalization now view a stronger state as a guarantor of democratic influence in the era of privatization, deregulation, and free market policy. The weakening of state authority under such measures suggests, therefore, that the involvement of civil-society forces is essential in building protections for social policy into global economic agreements. In this chapter, I will argue that certain structures of free trade agreements threaten to eclipse the function of civil society by undermining guarantees of transparency and participation by nongovernmental forces in the negotiation and implementation of these agreements. I shall argue also that, as a result of such trends, the normative task of civil society must be the introduction of "public spaces into state and economic institutions" (Cohen & Arato, 1994, p. 480), including multilateral decisions on trade and investment. "Democratizing the trade debate," Compa (1998) argues, "means conceding social justice advocates a serious participatory role in shaping the rules of economic integration and in the operations of the bodies that implement new rules" (p. 1). Without guarantees of public accountability[4] in trade agreements, Wallach (1993) argues, civil society forces face increasingly opaque barriers to environmental policymaking at the local and federal levels of the signatory nations.

I develop my argument in four sections. Initially, I identify the traits of civil society, including public participation, that appear to be the most vulnerable to an emerging neoliberal ideology guiding global integration of markets and capital investment. Second, I describe the rules under the 1994 North American Free Trade Agreement (NAFTA) for resolving disputes over trade and investment that restrict the participation of civil society forces. In the third section, I describe the proposed expansion of this model and its failures of transparency and public participation in the Free Trade Area of the Americas (FTAA) currently being negotiated by the United States and 33 other western hemisphere nations. Finally, I try to identify the challenges more generally that confront civil society groups in their effort to introduce "public spaces" into the emerging rules of economic globalization.

The description of the structures in trade agreements that restrict transparency and the participation of civil society groups in dispute resolutions is based on the text of the NAFTA agreement and the recently released (July 3, 2001), "bracketed" draft text of the FTAA.[5]

Civil Society and Economic Globalization

In his classic study of *The Structural Transformation of the Public Sphere,* Habermas (1991) argues that the early bourgeois public sphere was, itself, not unrelated to new developments in commercial trade and a corresponding need for news about markets and currency exchange (pp. 14–26). With the lessening of feudal power in the face of capitalist long-distance trade and the rise of the new modes of communication (the early "newsletters"), the conditions for civil society came into existence (p. 19). Occupying a central position within this sphere was a stratum of a bourgeois public to whom a new administrative state increasingly felt the need to address its attempts at regulation. The commanding status of this public "in the new sphere of civil society," Habermas argues, led inevitably to "a tension between 'town' and 'court,'" not unlike that in later modern society (p. 23).

Within the modern political state, civil society appears as "the ever-expanding array of organizations and networks that function independently of the state" (Burbach, 2001, p. 97). As such it has emerged as a sphere for discursivity and political struggle, and democratic accountability. In the United States, this sphere of public expression has achieved some form of institutionalization in statutes recognizing the rights of public participation, a "right to know," requirements for public notice of proposed regulations in the *Federal Register*, and, in certain agencies, the right of appeal.[6] And, it is this function of civil society (the mediation of state authority), I want to argue, that recent developments in economic globalization increasingly place in jeopardy.

Important to an understanding of the eclipse of civil society are new global institutions of trade and finance, particularly as these have been accompanied by what Falk (1999) has called "a group of ideas associated with the world picture of 'neo-liberalism'" (p. 1). Also referred to as the "Washington consensus," neoliberalism insists, especially in global arenas, on policies of "deregulation, privatization, structural adjustment programmes (SAPs) and limited government" (Held & McGrew, 2000, p. 5). Falk (1999) describes this policy consensus among elite leaders in the West as emerging in the aftermath of the collapse of the Soviet Union and the discrediting of socialism. In neoliberal discourse, he notes, "normative claims that insist on immediate and obligatory action by the state to overcome . . . poverty and joblessness are subordinated to a posture of deference to market forces and to a variety of economic restructuring priorities" (p. 100). Harnessed thus to an agenda of economic globalization, the discourse of neoliberalism is an ideological construction, "a 'necessary myth,' through which politicians and governments discipline their citizens to meet the requirements of the global marketplace" (Held & McGrew, 2000, p. 5).

Neoliberalism "discipline," however, has quite serious consequences for the regulatory role of individual nation states, and for the role of a public sphere. Neoliberal proponent Thomas Friedman (2000) argues that when a country "recognizes the rules of free trade in today's global economy," it puts on a "Golden Straightjacket" (p. 104). To fit into this straightjacket, he explains, a country must adopt certain "golden rules," including, among others, "opening its industries, stock and bond markets to direct foreign ownership and investment, deregulating its economy . . . [and] eliminating government . . . subsidies" (p. 105). Such rules discipline a state due to the fierce logic of the new, deregulated global economy. Brecher, Costello, and Smith (2000) explain, "the ability of governments to pursue . . . full employment or other economic goals[7] has been undermined by the growing ability of capital to simply pick up and leave" (pp. 8–9). As a result, Friedman says, the Golden Straightjacket narrows the policy choices of nation states to "relatively tight parameters" (p. 106). In such a view, the demands of civil society for transparency and public participation are viewed ipso facto as impediments to the free functioning and efficiency of markets, as a type of "*demosclerosis*" (Shamsie, 2000, p. 8).

The globalization of capital and neoliberal claims of a world market, Watts (2000) argues, "suggest that we are in the midst of a second 'great transformation' ([including] . . . the decomposition of civil society" (p. 143). At one level, elites view global institutions as a bulwark against the intrusion of "irrational" forces. One unnamed World Trade Organization official, for example, conceded, "The WTO is the place where governments collude in private against their domestic pressure groups" ("Network Guerillas," quoted in Brecher, Costello, & Smith 2000, p. 9). Such tendencies to shut the doors of multilateral regulation of trade thus threaten to marginalize the democratic forces emerging around international trade meetings. "In the main," Scholte (2000) observes, "global markets have seen power concentrated in what Robert Cox has evocatively called the *nébuleuse*, a process of élite networking that governs largely behind the scenes and free of public accountability" (p. 274, quoting Cox, 1992).

A somewhat more sophistical view also has begun to emerge, namely that electoral systems and government ministers subsume the role of civil society. Shamsie (2000) reports that one official representative at a trade conference went so far as to declare, "We are civil society" (p. 3). In this view, public participation is redundant and concepts of democratic participatory culture are seen as intrusions on or obstacles to market efficiency. Aune (2001) explains, "It is no accident that the quality of public argument has declined so dramatically in the years since the ascendancy of the market. The free-marketeers can only dismiss democratic participation as 'rent seeking behavior'" (p. 56).

The following sections seek to describe certain institutional means by which "free trade" agreements eclipse the demands of civil society for transparency, public participation, and democratic accountability.

BARRIERS TO PUBLIC PARTICIPATION IN NAFTA

Threats to the norms of public participation in nation states' social regulation became evident in the 1994 implementation of the North American Free Trade Agreement (NAFTA) by the United States, Mexico, and Canada. The trade agreement has been called the "most comprehensive free trade treaty ever signed," with its expansion of trade and investment rights for corporations (Dunkley, 2000, p. 89). Through restrictions against nontariff barriers (state and substate regulation of business operations), NAFTA has presumptively shifted a function of civil society—its mediation of state decisions—to nonpublic, administrative forums. Environmental critics have been especially concerned by this apparent weakening of the ability of citizen groups to influence environmental policy (Benton & Short, 1999, pp. 177–195). In this section, then, I describe two provisions of NAFTA that seem to pose the most troubling challenges to the sphere of civil society: (1) rules defining nontariff barriers and the indirect expropriation of corporate investment, and (2) procedures for resolution of trade disputes that are modeled after (nonpublic) commercial arbitration instruments.

NONTARIFF BARRIERS AND EXPROPRIATION OF INVESTMENT

So-called nontariff barriers to trade are, in neoliberal ideology, social regulations such as environmental policy, labor standards, or other restrictions on business operations that add costs to production or depress sales or anticipated income from sales. The authority for determining nontariff barriers is found in NAFTA's provisions for *investment* and investor-state relations. Parties under the trade agreement (Canada, the U.S., and Mexico) are prohibited not only from the direct expropriation of the property of a foreign corporation (investor), but also from taking any measure that "indirectly" may be "tantamount to nationalization or expropriation of such investment" (NAFTA, 1994, Ch. 11, Article 1110). This "Chapter 11" provision enables a corporation to sue a host government for actual or anticipated loss of profits (investment) resulting from environmental or labor regulations that constitute, in its view, an indirect *expropriation* by impacting business activities.

The most explicit example of a challenge under Chapter 11 on environmental grounds is the August 2000 decision in *Metalclad Corporation v. Mexico*. In this case, a U.S. firm, Metalclad, sought permits in 1994 to construct a hazardous waste landfill on property it had purchased in the state of San Luis Petosi, Mexico. Previously, the site had been used as a waste transfer station, where the facility had a history of polluting the local water supply. Local municipal officials, therefore, felt it was inappropriate as a site for a chemical waste landfill, given its location near an aquifer. After a geological audit showed the proposed landfill would contaminate the water supply, the state governor ordered the facility closed (Barlow, 2001a, p. 12). The Metalclad Corporation then sued Mexico for expropriation of its investment. The private NAFTA dispute tribunal hearing the case ruled against Mexico and awarded $16,685,000 to Metalclad (International Institute for Sustainable Development, 2001, pp. 74–79).

Another case involving a challenge to U.S. domestic environmental regulation has prompted intense criticism of Chapter 11 provisions in NAFTA. In 1999, the Methanex Corporation of British Columbia challenged the State of California's decision to phase out the use of MBTE, a gasoline additive. (The additive MTBE contains methanol, which Methanex manufactures.) DePalma (2001) reports that state officials have discovered that the gas additive, originally intended to reduce air pollution from motor vehicle emissions, is "a health hazard when it enters the water supply." Santa Monica, with 93,000 residents, he notes, "had to shut down most of its municipal wells when gasoline containing MTBE leached into the drinking water a few years ago" (p. 13). Foreseeing a loss of sales from the state's ban of MBTE, Methanex sued the U.S. government under in NAFTA's Chapter 11 provisions for the protection of investment, demanding $970 million in compensation. (The case is pending as of this writing.) As of April 2001, there have been 17 cases initiated under the provisions of Chapter 11. (For a summary of these cases, see International Institute for Sustainable Development, 2001, Annex 2.)

A party that loses a case before a dispute panel (see the following) must either accept the ruling (and, thus, change its practice or law) and offer appropriate compensation, or risk the retaliation by other parties of "benefits of equivalent effect" (NAFTA, 1994, Ch. 20, Article 2019). Such a presumptive shift of social regulation to the private arbitration of investor-to-state disputes effectively excludes the usual functions of civil society, that is, the capacity to deliberate and exert influence over the scope and direction of state authority. Andreas Lowenfeld, international trade authority at the New York University School of Law, observes, "There is no doubt that these [investment] measures represent an expansion of the rights of private enterprises vis-à-vis government" (quoted in DePalma, 2001, p. 13). Civil society groups have been more blunt. Some critics have charged that the cases filed thus far under NAFTA's "Dispute Settlement Procedures" showed how corporations were using the agreement "not to defend trade but to challenge the functioning of government" (DePalma, 2001, p. 13).

Nontransparency and Dispute Settlement

Private investor-state complaints about nontariff barriers under NAFTA are heard, not in the judicial systems of one of the signatory countries, but under nonpublic tribunals, whose authority has been drawn from commercial disputes under international law.[8] "These off-the-shelf mechanisms adopted by NAFTA have commonly been used to resolve private disputes between corporations, and are thus intended to provide a great degree of confidentiality" (DePalma, 2001, p. 1). The rules adopted from the U.N. Commission on International Trade Law for dispute settlement, for example, are the same as the procedures routinely written into private commercial contracts. Hence, Jernej Sekolec, Commission Secretary, observes that "arbitration [of investor disputes] is really private justice" (quoted in DePalma, 2001, p. 13).

The commercial context of these mechanisms helps to explain the absence in NAFTA's "Institutional Arrangements for Dispute Settlement Procedures" (1994, Chapter 20) of any requirement for public notice of trade disputes and rulings. The World Bank Center for Settlement of Investment Disputes, the second model used by NAFTA dispute panels, for example, is "bound by strict confidentiality rules, and only investors can say whether documents should be made public" (DePalma, 2001, p. 13). Under the Center's rules, the dispute proceedings can be opened to the public only if both the corporation bringing the complaint (the investor) and the government whose rules are being challenged agree.[9] The NAFTA proceedings, however, have never been opened to the public, DePalma reports, "nor have third parties until now been allowed to submit briefs.[10] Corporations want the proceedings to remain closed" (p. 13). Furthermore, the NAFTA tribunal decisions, technically, cannot be appealed.[11] Both proponents and critics of free trade agree that these provisions, drawn from other contexts involving essentially private commercial dispute settlement, "run headlong into demands for openness and accountability when public issues are involved" (DePalma, 2001, p. 1).

Officials from the World Bank and other trade offices appear to offer a consistent rationale for the closure of dispute proceedings to public—a need to preserve confidentiality to attract new corporate investment. "If increasing foreign investment is the prime goal in this [free trade agreements]," observes Ko-Yung Tung, Secretary of the World Bank's Center, "then making public these proceedings may be less important [than protecting investors]" (quoted in DePalma, 2001, p. 13).

Currently, the provisions of NAFTA regulate trade and foreign investment among commercial parties from the three parties to the agreement, the United States, Canada, and Mexico. This would change under the proposed Free Trade Area of the Americas (FTAA) being negotiated by the United

States and 33 other nations of the Western Hemisphere. In its current (draft) form, the FTAA agreement proposes to continue the current nontransparency and exclusion of civil society groups under NAFTA to provisions that would bind trade relations among these 34 nations.

Barriers to Transparency and Participation in the FTAA

Plans to negotiate the terms for a Free Trade Area of the Americas (FTAA) were officially initiated at the Summit of the Americas in Miami in 1994. The heads of state of the 34 nations of North, Central, and South America and the Caribbean (all except Cuba), agreed to extend many of the provisions of NAFTA to the region. The goal would be "to unite the economies of the Western Hemisphere into a single free trade arrangement" in which barriers to trade and investment were to be progressively eliminated (FTAA Web site). The FTAA negotiations were then launched formally in 1998, at the Second Summit of the Americas in Santiago, Chile.

At the Santiago summit, trade ministers established a Trade Negotiating Committee (TNC) to oversee the multiyear negotiations process and to ensure that "negotiations will be conducted in a transparent manner"[12] (FTAA Web site). Nine Negotiating Groups were set up to handle the major areas of negotiation among the 34 nations, including "Investment," "Market Access," and "Government Procurement."[13] Also established were three "special committees," including a precedent-setting Committee of Government Representatives on the Participation of Civil Society. Indeed, the FTAA Secretariat notes on its official Web site that "The FTAA is the first major trade negotiation where such a group has been established at the outset of the negotiations." And, it is the charge and the operations of the Committee on Civil Society that I examine below.

Nontransparency in FTAA Negotiations

Early prospects seemed encouraging for the recognition of the importance of civil society in hemispheric governance. Indeed, the Summit of the Americas' Action Plan in 1994 called for mechanisms to expand citizen participation generally (see Shamsie, 2000, pp. 17–18). And the Office of the U.S. Trade Representative echoed the importance of transparency and participation of civil society in a statement published in the *Federal Register* (Office of U.S. Trade Representative, 1998).

Discussions of a new trade area at the 1998 Summit of the Americas in Santiago, however, occurred in a very changed context for the recognition of civil society. Shamsie notes that a proposal for including civil society groups directly in the FTAA negotiating process "never made it through the preparatory meetings" for the Summit (2000, p. 21). Instead, FTAA trade ministers, after what was termed "intense behind-the-scenes haggling," established a Committee of Government Representatives on the Participation of Civil Society (CGR) with the mandate of "encouraging" civil society groups to provide their views on trade issues (Shamsie, 2000, p. 14). This compromise seemed to reflect renewed skepticism in Santiago by many ministers toward recognizing a role for civil society. One official commented, "Even saying the words 'civil society' in a trade forum brings all kinds of wrath upon you" (quoted in Shamsie, 2000, p, 14).

The workings of the nine Negotiating Groups, which meet periodically, seem to bear out trade ministers' skepticism of civil society groups. Since 1998, the Groups have proceeded to work in secrecy with no release of its documents or a draft text of the FTAA until July 3, 2001. Over 500 corporate representatives, nevertheless, have been given security clearance and access to these negotiating documents (Barlow, 2001a, p. 4). In terms of direct access, Shamsie (2000) notes that members of a private business advocacy group, the Americas Business Forum, "have far greater access to trade and finance ministers than do any of the participants in the parallel civil society processes" (p.10). Many civil society groups have, therefore, called for the release of the relevant negotiating documents and drafts of the treaty itself.[14] Others have formed a strategy of trying to meet with or pressure their own government's representatives to the FTAA process to share information about the negotiations.[15]

Concerns for transparency have extended to members of the U.S. Congress as well. Most members have been excluded from security clearance and, thus, prevented from viewing the official texts or learning the names of the U.S. representatives to the Negotiating Groups. On January 5, 2001, a bipartisan group of members wrote to President Clinton, requesting that his Administration "impose a simple standard of openness and public accountability in the negotiations on creating a Free Trade Area of the Americas" (Rep. Peter DeFazio, personal communication). In response to this request, the Office of the U.S. Trade Representative (USTR) released a brief summary of its negotiating positions. Again, in March 2001, the same members requested from President George W. Bush that he follow through on their initial request to President Clinton, by:

- Immediately publishing all working papers, memoranda, and other text, including the bracketed text of the actual FTAA;
- Providing members of Congress with a list of the U.S. representatives to the nine working groups to promote a more direct dialogue; and
- Making available to nongovernmental organizations and the American public through the USTR public reading room and on the USTR's Web site all of the aforementioned materials. (Rep. Peter DeFazio, personal communication, March 15, 2001)

One U.S. nongovernmental group, Public Citizen, has charged that the Office of the USTR violated several laws requiring disclosure, including the Administrative Procedure Act, Government in the Sunshine Act, and the Federal Advisory Committee Act (Public Citizen, 1998, p. 1). When the USTR office refused to release the draft FTAA text publicly, the nonprofit law firm Earth Justice Legal Defense Fund filed a lawsuit under the Freedom of Information Act to force release of the document (Palmer, 2001).

Finally, at the 2001 Summit of the Americas in Quebec City, Canada, the 34 heads of state agreed to release publicly a version of the FTAA text. After a short delay, the FTAA released a "heavily edited 430-page draft" of the agreement on its official Web site on July 3, 2001 (Dunne, 2001, p. 4). Civil society groups remained dissatisfied. Lori Wallach of the nongovernmental group Public Citizen described the "bracketed" text[16] released as "a fragment of the total agreement, one that has been sanitized by eliminating vital information" (Dunne, 2001, p. 2). As a result, U.S. Representative Lloyd Doggett complained that the administration had shown "only minimal interest in . . . assuring a reasonable level of transparency and public participation in trade decision-making" (quoted in Dunne, 2001, p. 4).

BARRIERS TO THE PARTICIPATION OF CIVIL SOCIETY

The FTAA trade ministers had intended that the special Committee of Government Representatives on the Participation of Civil Society (CGR) would be the route through which civil society groups would interact with the negotiating process. The CGR appears, at first glimpse, to have a straightforward mission and rationale for the contributions of civil society. Its mandate, developed by the trade ministers at the 1998 San José Ministerial meeting, is "to receive inputs from civil society, to analyze them and present the range of views to the FTAA Trade Ministers" (FTAA Web site). The mandate notes the statement of Trade Ministers on civil society: "We recognize and welcome the interests and concerns that different sectors of society have expressed in relation to the FTAA" (FTAA Web site).

The CGR, however, was slow to develop a work agenda for itself and a mechanism for consultation with civil society groups. Shamsie (2000) notes that the FTAA governments "watered down" the initial mandate, charging the CGR only to transmit the views of civil society organizations to the Negotiating Groups and only on narrowly defined "trade-related matters" (pp. 14, 15). Although the other two Special Committees (Small Economies and Electronic Commerce) were authorized "to make recommendations," the CGR was allowed only to "present the range of views" of civil society to trade ministers (p. 15). Before the first meeting of the CGR in 1998, a large number of civil society groups drafted a letter to the CGR requesting access to information and opportunities to discuss negotiations. The letter closed, Shamsie reports, by arguing that "real participation could occur only if the opinions expressed were transmitted in a way that could potentially influence the outcome [of] the negotiations" (p. 15).

In 1999, the CGR unveiled the committee's official route for consultation with civil society groups, termed the "Open Invitation to Civil Society." This mechanism was an invitation to these groups to submit their views to the FTAA process on "trade-related" issues. Interested groups were given five months to respond. The publicity for the "Open Invitation" was left to each government to initiate; although some governments did publicize the invitation, most did not. (Shamsie, 2000, p. 15).

Most civil society groups were disappointed with the CGR's "Open Invitation" proposal. One civil society representative from the United States observed that it was difficult to convince many Latin American civil society groups to submit their views. "Without the establishment of a direct link to the negotiating groups, reasonable timeframes for consultation, and clear procedures of accountability," Feinberg and Rosenberg (1999) report, "the ['Open Invitation'] initiative could become only a repository for [civil society] concerns with no chance of influencing the negotiating process" (p. 688, quoted in Shamsie, 2000, p. 15). In November 2000, the Hemispheric Social Alliance, on behalf of more than 300 groups in the hemisphere, wrote to Dr. Adalberto Rodriguez Giavarini, the Chair of the FTAA Trade Negotiating Committee, demanding the text of the FTAA be publicly released and noting deficiencies in the invitation to provide input to the CGR. The Alliance noted that even if the groups were to make submissions to the CGR, "the result is not the participation of civil society in this process but simply a one-way communication. It is impossible for us to engage in a serious dialogue on the FTAA when we do not know the actual content of the negotiation" (Hemispheric Social Alliance, 2000, p. 1).

As a result, many civil society groups have expressed dissatisfaction with the ability of the Committee of Government Representatives to incorporate the views of nongovernmental groups and noncorporate interests. Public Citizen, a U.S. nonprofit organization concerned with the democratization of trade rules, used the "Open Invitation" more broadly to submit recommendations for ensuring public input into the FTAA process. In its written comments, the group recommended that the CGR must proactively seek public input from consumer and environmental organizations and other representatives of civil society. Specific recommendations included: (1) the CGR should extend the openness and public participation requirements of U.S. law to the FTAA process. Procedural transparency and access to information should be the rule for the CGR's agenda, meetings, and reports as well as for the agenda, meetings, and reports of the other FTAA negotiating groups and the Trade Negotiations Committee and (2) the CGR should produce recommendations for the negotiating groups on steps that should be taken to address the issues raised by citizen groups (Public Citizen comments, 1998).

In the end, only 77 comments were submitted in response to the "Open Invitation," the majority from business-related nongovernment organizations. The CGR Web site provides an "Executive Summary" of the comments submitted but with no clear account of the distribution of, consideration by, or the influence of these comments on the Negotiating Committees. Indeed, one FTAA official expressed fear that even sending the "Open Invitation to Civil Society" would flood the CGR with civil society views. "The same official noted that this concern explained why no plan was developed to deal with the received submissions" (Shamsie, 2000, p. 8). As a consequence, Maude Barlow of the International Forum on Globalization has argued that "the Committee's real role is not to listen, but to keep up the appearance of real dialogue" (2001a, p. 21). Some support for this view came from an unexpected source. Sherri Stephenson, Deputy Director for Trade for the Organization of American States, one of the three international bodies assisting the FTAA, observed that the benefit of the committee's work "may diffuse pressures related to issues of labor and the environment" (quoted in Barlow, 2001a, p. 21).

The failures of transparency and participation by civil society groups in the FTAA negotiations also are reflected in drafts of the actual trade agreement itself. The July 2001 release of the draft (bracketed) FTAA text discloses the extension of some of the features of nontransparency and absence of meaningful public participation in NAFTA.

Nontransparency and Participation in Dispute Settlements

It is unclear from the released draft text whether the FTAA negotiating parties will adopt the NAFTA model for investor-state dispute settlement or an even stronger World Trade Organization provision for a state-to-state dispute settlement system.[17] Under the latter (WTO) model, a country, often acting on behalf of a domestic corporation of that nation, can directly challenge an existing law or policy (e.g., an environmental or labor regulation) of another country. A country that loses in such dispute proceedings has three choices: "change its law to conform to the WTO ruling; pay permanent cash compensation to the winning country; or face harsh, permanent trade sanctions from the winning country" (Barlow, 2001a, p. 20). Barlow believes that FTAA negotiators will choose ultimately "to retain the powers of private dispute settlements contained in the investor-to-state provisions of NAFTA, while opting for the more stringent conditions of the WTO to settle state-to-state disputes" (pp. 20–21). Under either model, civil society groups' interest in transparency and public participation appear to be foreclosed. Both NAFTA and WTO rules for dispute resolution deny a right of notification to the public, maintain the confidentiality (secrecy) of dispute proceedings, and have traditionally prohibited intervention by third parties.

In summary, since the launch of the FTAA negotiations, civil society groups have expressed disappointment and frustration at the lack of openness and the exclusion of meaningful public participation in the FTAA process. It remains to be seen whether their criticism of the procedures adopted by the trade ministers for inclusion of civil society views will influence future negotiations or be incorporated into new trade agreements.

CHALLENGES TO OPENNESS AND PARTICIPATION IN "FREE TRADE"

Barlow (2001b) notes, accurately, I believe, that "the only way we are going to re-claim and democratize the state is with an active civil society engaged in real participatory democracy" (p. 4). But how? How can civil society pursue a normative project to introduce "public spaces into state and economic institutions" (Cohen & Arato, 1994, p. 480), including, most importantly, multilateral decisions on trade?

In the aftermath of antiglobalization protests at the ministerial meeting of the World Trade Organization in Seattle in 1999, many social groups have pursued a strategy of public confrontation and convening of alternate global forums paralleling these meetings. A robust civil society now regularly gathers in connection with international meetings of World Bank, IMF, the Summits of the Americas, and G-8 conferences, as well as free trade meetings. Pieterse (2000) argues that, as a result, a "collective awareness of concerns . . . such as the environment, population [and] development—has been growing, and so has its public articulation, notably in U.N. global conferences, so that arguably, a global public sphere is emerging" (pp. 1–2). Indeed, some progress has been achieved in creating channels for greater input from citizen groups in some international bodies. Compa (1998) cites recent reforms in the European Union (EU) and World Bank. These include the creation of trade union, environmental, and other nongovernmental groups' advisory bodies in the EU and the new Structural Adjustment Participatory Review Initiative (SAPRI) in the World Bank to consider the views of civil society (pp. 1–2; cf. also Scholte, 2000, p. 271).

Still, the creation of advisory councils or other consultative arrangements are, by themselves, insufficient to "reclaim and democratize" global trade or investment rules. Although well intentioned, Compa notes, "these mechanisms usually involve decision-makers telling their interlocutors, 'We'll listen to what you have to say, and then we'll do what we want to do'" (p. 2). It is important, therefore, not to exaggerate the influence or degree of actual participation and mediation of multistate authority these have provided. Pieterse (2000) cautions that such extrainstitutional modes of expression and awareness of the "dark side" of globalization "do not 'line up' or add up to a condition of collective capacity" (p. 2).

Important, too, is the creation of meaningful "public spaces" inside global institutions for trade that allow for consideration of divergent views, access to decision-making bodies, feedback, and accountability to wider, democratic constituencies. Certainly an outline already exists for reforms of this nature. After the 1998 Summit of the Americas in Santiago, for example, the nongovernmental group Foreign Policy in Focus made four recommendations to trade ministers, calling for civil society groups to have:

- Parity with financial and corporate interests in access to the policy planning process;

- Access to negotiating forums and procedures where trade and investment rules are drawn up and new institutions are established;
- Access to enforcement and dispute resolution mechanisms that implement new rules; and
- Equal consideration for staff positions in new regulatory positions. (Compa, 1998, p. 4)

Along with such suggestions offered in response to the "Open Invitation" during the FTAA negotiating process by Public Citizen (perviously mentioned), the basis for achieving institutional reform in the workings of trade and investment rules is not technically difficult.

What is becoming clear, however, is that, alongside proposals for specific institutional redress, other, more discursive challenges also must be addressed. Two questions in particular have been put to civil society representatives by ministers at free trade conferences: (1) "Whom do civil society groups speak for?" and (2) "What alternatives do such groups propose to neo-liberalism's logic of market discipline?" These challenges—of representation and articulation of an alternate vision—remain at the center of efforts by civil society groups to open more meaningful spaces within the institutions of global trade. Let me consider these briefly in closing.

The problem of the representation of civil society groups arises frequently in the exchanges between officials and nongovernment groups in international contexts. Demanding a greater voice in trade forums, Compa (1998) argues, carries a certain responsibility for civil society. "A threshold issue is determining who speaks for social justice [including environmental] communities in policy debates and in negotiations over new, more democratic rules for trade" (p. 3). In her examination of the role of civil society in three western hemispheric organizations—the Summits of the Americas, the FTAA process, and the Organization of American States (OAS)—Shamsie (2000) found that government officials demanded to know: "Who do civil society organizations represent?" and "How are civil society organizations held accountable?" (p. 4) She noted that governments are hesitant to bring civil society groups into consultation "because it is difficult to determine for whom they speak" (p. 4).

In the western hemisphere, Compa notes, the formation of the Inter-hemispheric Regional Workers Organization (ORIT) has been recognized by its labor members as the body to speak for unions in trade forums because the organization embraces nearly all labor federations in the region (1998, p. 3). But other civil society forces face a formidable challenge. There is no ORIT for environmental, human rights, or other groups. As a result, Compa argues, "to have an effective voice in trade policy debates, NGOs must be able to answer the question, 'Who do you represent?'" (p. 3). On the other hand, if the aim of consultation is to hear from points of view that diverge from governmental consensus, then, as some argue, the participation of a robust civil society may be justified apart from issues of representation and accountability (Shamsie, 2000, p. 4).

Finally, there is also a need for civil society forces to be able to articulate persuasive alternatives to the current rules of economic globalization and its neoliberal rationale. It is no longer enough for demonstrators to chant, "Just say 'No' to the WTO," although such chants will continue to mobilize popular support for the movement. The voicing of a vision of global relations that nurture a just and sustainable economy, as a counterdiscourse to neoliberalism's "Golden Straightjacket," is also needed. And, as is evident from recent efforts—from the People's Summit outside the garrisoned meeting halls of the heads of state at the Summit of the Americas in Quebec City to the recent work of antiglobal scholars—such a vision is beginning to emerge. Whether termed "globalization-from-below" (Falk, 1999) or "democratically governed market economies" (Korten, 1999), a consensus seems to be arising around a set of normative ideas. Falk, for example, has offered a persuasive summary of the elements of a new, "normative democracy" able to incorporate the consent of citizens, participation, accountability, and transparency into the rules and institutions regulating global markets and finance (1999, pp. 148–149).

The ability of civil society forces to persuade states to open meaningful channels for public participation in trade forums is, arguably, dependent on resolving these two challenges—the development of a compelling rationale for the inclusion of environmental, labor, and social justice voices in trade negotiations, and a compelling vision of the democratization of trade rules. Ultimately, the struggle to democratize free trade and investment is not just about expressing a diverse view, but about mediating state (and multistate) power, the historical rationale for the emergence of a sphere of civil society. A mature vision of a more democratic globalization may not yet be at hand, but as Falk (1999) reminds us, its "human prospects depend on struggle, resistance, and vision," and such resistance is "best guided by an attuned, if diverse, embryonic global civil society" (p. 33).

NOTES

1. Cohen and Arato (1994) note that contemporary conceptions of the term generally agree that "civil society represents a sphere other than and even opposed to the state. All include . . . some combination of networks of legal protection, voluntary associations, and forms of independent public expression" (p. 74).

2. The shift in emphasis emerging as a result of the seven-year Uruguay Round of trade negotiations is generally seen as the most dramatic step that governments have taken collectively in the direction of free trade between nations (Dunkley, 2000, p. 3). For a detailed analysis of the new disciplines agreed to in the Uruguay Round, see especially Dunkley (2000) and Wallach (1993).

3. For this argument, see Scholte (2000), pp. 274–275.

4. Public Citizen, a leading civil society group concerned with trade, insists that "accountability" in trade agreements must include "the rights of notice, comment, the opportunity to participate, the ability to bring complaints, and access to decision-making processes" (Wallach, 1993, p. 43).

5. The draft (bracketed) text of the proposed FTAA is available on the official Web site of the FTAA Trade Negotiating Committee at: www.ftaa-alca.org/alca_e.asp

6. The principle of public participation in matters of environmental policy, particularly, is inscribed in the precedent-setting National Environmental Policy Act (NEPA) and, more recently, in the federal rules promulgated under President Bill Clinton's 1994 Executive Order on Environmental Justice. The regulations implementing the requirements of NEPA for public participation, for example, require federal agencies to "request the comments from . . . those persons or organizations who may be interested or affected" ("Regulations," 1978, Section 1503.1, 43 Federal Register 55997).

7. For impacts on environmental policy, see Alliance for Responsible Trade (n.d.).

8. Resolution of investor-state disputes under NAFTA draw from one of two procedures, either from the World Bank's International Center for Settlement of Investment Disputes or from the U.N. Commission on International Trade Law, based in Vienna, Austria (DePalma, 2001).

9. Some U.S. nongovernmental groups have use of requests under the Freedom of Information Act to force the U.S. government to release documents when it has been named a defendant in a dispute settlement proceeding (DePalma, 2001, p. 13).

10. The reasons for this are unclear. NAFTA does contain a relevant provision: "A Party that is not a disputing Party, on delivery of a written notice, shall be entitled to attend all hearings [and] to make oral and written submissions to the panel" (NAFTA, 1994, Chapter 20, Article 2013). Earth Justice Legal Defense

Fund, for example, has asked the tribunal that is hearing the Methanex case "to consider breaking with tradition and accepting written arguments from third party groups like the Bluewater Network, a citizen's environmental group" (DePalma, 2001, p. 13). DePalma reports that the tribunal has determined that it has the right to accept such written arguments but has delayed a decision on whether it will do so in this case (p. 13). See *Methanex v. United States*, Decision of the Tribunal on Petitions of Third Parties to Intervene as "Amici Curiae" [On-line]. Available: www.iisd.org/pdf/methanex_tribunal_first_amicus_decision.pdf

11. NAFTA tribunal rulings may only be reviewed by a local court for corruption or gross misinterpretation of the rules by the hearing panel (DePalma, 2001, p. 13).

12. The mandate of the FTAA's Negotiating Group on Dispute Settlement includes the injunction, "to establish a fair, transparent and effective mechanism" among FTAA countries. "Transparency" in this context, nevertheless, appears to refer to business-state protocol (e.g., the right of investors to full knowledge of required permits or other regulations affecting investment), rather than to a public disclosure requirement generally. See http://www.ftaa-alca.org/ngroups/ngdisp_e.asp

13. The nine Negotiating Committees cover the areas of Market Access, Investment, Services, Government Procurement, Agriculture, Dispute Settlement, Intellectual Property Rights, Subsidies, Antidumping, and Countervailing Duties, and Competition Policy.

14. An early draft of a report of the FTAA Negotiating Group on Investment was leaked and, subsequently, briefly made available to the public on the Web site of the Institute for Agriculture and Trade Policy. The URL, which is no longer active, was www.iatp.org/foodsec/library/admin/FTAA_Negotiating_Chapter_Made_Public.html

15. See Shamsie (2000) for a brief description of a 1999 meeting of 20 trade ministers with some nonbusiness civil society organizations in Toronto (pp. 10, 23, fn. 4).

16. The term *bracketed*, in this context, refers to the passages of the FTAA draft text that are in dispute by one or more nations. The full draft document would normally include these competing proposals or interpretations of the text, but this material has been deleted from the version of the FTAA released to the public in July 2001.

17. The Negotiating Group's mandate instructs negotiators to take "into account inter alia the WTO Understanding on Rules and Procedures Governing the Settlement of Disputes." See http://www.ftaa-alca.org/ngroups/ngdisp_e.asp

Part Three

Emergent Participation Practices Among Activist Communities

CHAPTER 11

GLOBAL GOVERNANCE AND SOCIAL CAPITAL: MAPPING NGO CAPACITIES IN DIFFERENT INSTITUTIONAL CONTEXTS

AMOS TEVELOW

Bolstered by an ideology and marketing strategy developed in policy networks throughout the 1970s, neoclassical economics swelled during the 1980s, as Ronald Reagan and Margaret Thatcher pursued structural adjustment policies[1] through the International Monetary Fund (IMF) and World Bank (WB). When communism fell, euphoric voices announced that history ended.[2] The Clinton-led ascent of the Democratic Party on a pro-NAFTA antiwelfare platform, modulated by a "Third Way," also showed the rhetorical advantages of market piety. But cracks in the "end of history" thesis began to appear in the late 1990s as elite and popular challenges to global capitalism converged in opposition to a pure free market in trade, finance, investment, and development—represented in terms of a "Washington consensus." The 1997 "Asian flu" led even hard-core neoclassical economists to seriously reconsider of the IMF's headlong dive into financial deregulation while, simultaneously, broad coalitions of environmental, labor, and indigenous activists mobilized to defeat the 1998 Multilateral Agreement on Investment, shut down the 1999 World Trade Organization (WTO) meeting of ministers from its 134 member countries in Seattle, and alter the WB's agenda for dams, thermal plants, natural resources, debt relief, AIDS, and landmines. The confluence of these tendencies constitutes the first post–Cold War "legitimacy crisis" for the global governance complex.

This chapter examines several cases to understand how specific features of global governance institutions such as official participation mechanisms, status of sympathetic insiders, and issue focus affect nongovernmental organization (NGO) challenges to the governance complex. It also points to *social capital* as an independent variable affecting NGO capacities. Social capital is an important condition for effective NGO campaigns, but has serious shortcomings as a policy model. For sociologists, social capital refers generally to advantages gained by being active in networks that solidify trust and shared norms over time (Bourdieu, 1986; Coleman, 1990; Granovetter, 1985; Putnam, 1993). Transnational coalitions built on norms of trust and mutual influence are able to respond quickly and cohesively through the Internet. Investing in interorganizational chains across differences in culture, geography, and economic and political power appears to be essential to productive coalitions. A small number of people connected together in a global public policy network can exert power disproportionate to their resources only by bridging major gaps in culture, power, and resources.

However, social capital used as a policy tool may actually prevent more fundamental reforms by "humanizing" governance without rigorously and explicitly addressing issues of equity and justice. As normative concerns, equity and justice were hardly countenanced by the Washington consensus, so even addressing them obliquely through buzzwords like governance, capacity building, social capital, institution building, safety nets, and so on, represents a subtle improvement on neoliberal conceptions of market dynamics. But this small gain might be outweighed if orthodox economics becomes applicable to previously untapped areas of the social. Impoverished neighborhoods, cities, regions, and countries may be expected to converge not just on the economic norms of structural adjustment, but now also on a set of sociopolitical norms imposed by the WB. Good government will entail the manipulation of public participation and democratic accountability as variables toward efficient and effective outcomes (Reinecke, 1998). The WB's use of social capital as a development concept shows the extent to which governance can sidestep issues of power and overcome resistance.

ORGANIZATION FOR ECONOMIC COOPERATION AND DEVELOPMENT (OECD)

Objections from developing countries prevented rules for direct foreign investment from being included in the postwar General Agreement on Tariffs and Trade (GATT), and at its renegotiation in the Uruguay Round in the

1990s. After failing to incorporate direct investment rules in 1995, industrialized states shifted discussions to a friendlier institutional context, the Organization for Economic Cooperation and Development (OECD), which had been studying and preparing for negotiations to liberalize investment rules since 1991 (Tieleman, 2000, p. 8; Kobrin, 1998).

But a Multilateral Agreement on Investment (MAI) never came. Negotiations were scuttled toward the end of 1998 after delays caused by intense public scrutiny, and are not likely to be renewed. The international campaign to defeat the MAI exemplifies the ways that heterogeneous NGOs in different countries can draw on and build bonds of trust through social mobilization activities based on shared values. A draft text of the MAI was leaked in February 1997. A network of NGOs took informal leadership positions and quickly distributed the text with analysis through e-mail lists.[3] Feeling purposely excluded from negotiations, and outraged at the antidemocratic terms of agreement, a coalition of over 600 NGOs banded together with sympathetic politicians and trade unions, and dedicated themselves to halting the MAI negotiations by publicly exposing its flaws (Tieleman, 2000, pp. 8–15).

One explanation of the MAI failure is an accumulated social capital deficit between the OECD and civil society due to OECD structural consultation procedures. Its 1960 convention mandates that OECD act more like a think tank to promote "the further expansion of world trade" (OECD, 1960). OECD has a limited tradition of negotiation, and no official role for NGOs. An independent Negotiating Group bypassed any ties that environmental, fair trade, and human rights NGOs might have had to OECD committees. The Negotiating Group was a homogeneous group of investment experts, policymakers, and diplomats sent by the member countries and business and trade union representatives who, as Dutch assistant chair Jan Huner (1998) observed, "were not used to viewing from a political perspective the concepts that they consider logical and essential parts of an investment discipline. . . ." (p.1). Seeing their actions as largely technical in nature, the negotiators took the benefits of the treaty for granted, and were genuinely surprised by controversy. They also clearly underestimated NGOs' capacities to coordinate an international campaign to educate and mobilize citizens and public officials.

By the time negotiators invited NGO participation in late October, mistrust was so deep that most decided to fight the MAI instead of consult. Even though the OECD signaled that environmental and labor considerations could for the first time be made as part of an investment treaty, NGOs felt that their involvement would merely legitimate a predecided agenda. When solicited for a seminar with business and labor a year later (December 2, 1998), many in the NGO network were not even on speaking terms with OECD negotiators and rejected the meeting on the grounds that it was designed to split the movement. They reasoned that since the OECD had seen fit to meet with transnational corporations rather than civil society groups for two years, they couldn't be trusted to take NGO concerns seriously now (Tieleman 2000, p. 16). Had the OECD respected the competencies of NGOs, they might have channeled regular and accessible information back and forth in a continuous accountability system. Instead, NGOs were not invited to negotiations, leading to a spiral of mistrust and a campaign to "kill the MAI." When the MAI was finally pulled, activists hailed it as a victory of millions of Web-based Davids versus Goliath, and it became a representative anecdote of a sort of ideal form of public participation characterized by publicity and information-sharing activities of informally linked NGO networks, representing diverse geographic zones, to achieve legislative victories.

INTERNATIONAL MONETARY FUND (IMF)

Rather than take responsibility for creating the conditions underlying the 1997 East Asian financial crisis, IMF leadership tried to spin the Asian crisis as a rebuke to "crony capitalism." But when IMF solutions seemed to exacerbate the problem, economists in elite forums decried the excesses that external critics had been voicing for twenty years (and internal critics, famously WB Chief Economist Joseph Stiglitz, more recently). The failed IMF response led to a genuine debate among mainstream economists on the ability of the free market to mitigate the dislocations resulting from globalization (see Krugman, 1999; Rodrik, 1997). The notion that free markets and reduced government would not work if not embedded in civil society institutions began to emerge.

In policy communities there is now a general consensus about the need for a "Post-Washington Consensus" (PWC) (Stiglitz, 1998), a new development paradigm that takes a strict separation of state and market to be analytically and practically problematic. It sounds good on paper—social theorists, third-world activists, and Keynesian economists had been saying so for years. But (1) Although Stiglitz and others did represent a significant departure from the narrowly technical models for decision making in place, they fully embraced the principle of opening markets, and (2) Even this moderately progressive position was quashed when Stiglitz was fired after making one too many uncouth statements about the IMF. When the wording of the World Development Report was too radical for U.S. Treasury Secretary Larry Summers, head author Ravi Kanbur was also pressured to resign. Rather than taking big steps to rescind structural adjustment policies, or include NGOs (Alexander, 1999; Sachs, 1997; Scholte, 1998), the governance complex has preferred to expel dissidents.

Many in the global justice movement, as well as nativists of the Pat Buchanan variety, responded to the crisis with calls to withdraw from global governance institutions altogether. But abolitionist positions did not prevail among policymakers, and currently two mainstream versions of the PWC vie for control. One is unreconstructed neoclassicism—the Bush administration position, readable in conservative think tanks such as the Heritage Foundation, and the majority of the Congressionally appointed and highly critical Meltzer Commission. They support the IMF, WB, and WTO only insofar as they deregulate and privatize, and are wary of using the IMF or WB as aid agencies for poverty alleviation and financial bailouts (Judis, 2000).

The other is neoliberal internationalism—Treasury Secretary Larry Summers, World Bank President James Wolfensohn, Thomas Friedman, and economists at the Institute for International Economics and Brookings Institution. They acknowledge limits to market solutions but accept the basic tenets of neoliberalism. They urge modest reforms in global institutions such as more "transparency," more cautious prescription of high interest rates at IMF, and a greater focus on poverty and AIDS at the WB, but oppose considering environmental and labor standards in trade treaties, and support loans with structural adjustment conditions attached (Judis, 2000).

Although Stiglitz's moderately progressive position has become a marginal one in the emerging PWC, how did a view so out of whack with the dominant neoliberal paradigm make it so far? How did the WB, of all the three most powerful global governance institutions (WTO, IMF, WB) become the most self-reflective? Consider the uses of social capital at the WB.

WORLD BANK

Twenty years of NGO activism has been a major factor impacting World Bank projects (e.g., dams, highways, thermal plants, natural resources) and systematic reforms (e.g., public information policies, structural adjustment conditions, inspection panels, joint evaluations, and NGO liaisons) (Chiriboga, 2000). The popular pressure exerted by the NGO networks throughout the 1990s correlates with increasing levels of NGO involvement in multilateral debt and development talks at the Bank. Where NGO involvement in the WB's lending projects averaged six percent between 1973 and 1988, by 1998 about half of the lending projects provided for NGO involvement (Simmons, 1998, p. 3). When the aim is to control project damage or shape local implementation, opposition tends to be spearheaded by grassroots organizations directly affected by WB projects. When the main concern is systemic reform of bank policies and institutional arrangements, transnational NGO issues networks exert sustained pressure and benefit from advice and information offered by sympathetic reformers within the WB (Brown & Fox, 2000, pp. 2–3).[4] Capacities for future NGO campaigns are thus enhanced. Internal reformers also benefit from the activities of external pressure groups, which threaten reluctant WB members with embarrassment by exposing drafts through early publication. This sort of reciprocal information sharing is evidence for the presence of social capital.

Over the last 10 years, 150 public interest NGOs have collaborated in the Fifty Years is Enough campaign for greater transparency and responsiveness at the WB and IMF. Hundreds of NGOs from all continents focused attention on the harmful actions and secrecy of the institutions. By publicizing debt and development issues, the network exacted significant concessions from the WB, including:

Debt relief. Rooted in econometric analyses showing development has been arrested and even reversed by the North's domination over the South (Bello, 2001; Weisbrot, Naiman, & Kim, 2000), third-world NGOs focused on debt alleviation and equitable development strategies as reparations for the damage the IMF and World Bank have caused in Africa and the South. With some success they have vigorously promoted the idea that, like credit card debt that can never be paid back because the interest to service is too high, the 2.2-trillion-dollar (in 1999) third-world debt is an unconscionable burden to humane economic development, especially because corrupt authoritarian governments accrued much of it. NGOs have argued that although IMF and WB demands that third-world countries pay down debt by earning hard currency through exports, money earned through export-led growth should go more toward domestic investment than international collection agencies. The IMF and WB conceded the need for limited, but not significant, debt reduction, and has attached the same structural adjustment strings (opening up to foreign investment and trade, export-led development, etc.).

Transparency. The Narmada Campaign around the Sardar Sarovar Dam Project shows how mobilization for a project turned into a longer and larger mobilization to change central WB policies regarding public information and participation (Chiriboga, 2000). In 1990, a partnership of local and international NGOs rallied against the WB's role in constructing the Sardar Sarovar Dam on the Narmada River in western India. An Internet alliance of fewer than 20 people pushed for a policy of early access to WB information and an inspection panel to investigate project damage. World Bank staff in favor of a more open disclosure policy supported the struggle, as did U.S. Congress members, who threatened to withdraw future funds if the WB did not comply. The WB pulled out in 1991, setting the stage for future reforms. An independent inspection panel was created in 1994 to provide an impartial forum for board members and citizens to protest bank projects (World Bank, 1994). In contrast to the WTO, the World Bank in the 1990s began to readily disclose many documents. Further steps to dialogue with NGOs were taken when James Wolfensohn became president of the World Bank in 1995. Wolfensohn has also moved the WB away from massive projects like hydroelectric dams toward education and small-business credit-assistance programs.

But the World Bank is an interesting case not just because it has been more reflective and reformist in the face of NGO pressure than the WTO and IMF, but also because it has deployed the notion of social capital to incorporate (and deflect) criticism. The WB maintains a Web site dedicated to the notion of social capital (http://worldbank.org/poverty/scapital). The site is organized around sources of social capital (family, communities, civil society) and topics such as economics, trade, finance, health, nutrition, technology, sanitation, water supply, and urban development. Nowhere is social capital used to critique existing structural adjustment policies, or call into question basic economic assumptions at the World Bank.

As Fine (2001) points out, social capital rhetoric at the WB represents just a minor tweaking of economic theory, and functions to broaden the scope of intervention into nonmarket spheres. The PWC is couched in language friendly to state and social institutions, making it palatable to both clients and critics (Phillips & Higgott, 1999). Social theorists once marginalized at the WB a rush of legitimacy, but the PWC opening is small indeed, and the newfound sense of acceptance among sociologists and anthropologists should not be mistaken for a substantive challenge to the economists.[5]

WORLD TRADE ORGANIZATION

Nongovernmental organizations saw the mobilization against the MAI as an auspicious proving ground for the 1999 WTO meeting in Seattle. But despite the visibility of the Seattle protests, WTO has not been subject to as much internal criticism or actual reform as have the other two Bretton Woods institutions (IMF, WB), or the U.N. This is both an effect and cause of WTO's relative isolation from NGO actors, which has produced a significant lack of trust. NGOs in the "Fifty Years is Enough" network view the WTO as a pure extension of IMF/WB imperatives of privatization and deregulation into trade, making the imperatives prerequisites for entry into the world market (as well as conditions for IMF/WB loans). The Free Trade Agreement of the Americas embodies the market fundamentalism found in the IMF, WB, and WTO. It is the latest codification of similar structural adjustment rules and enforcement mechanisms, allowing corporations, for example, to sue governments that impose environmental regulations on industries.

WTO efforts to solicit limited NGO input into trade deliberations can also be seen in the context of "good governance" and "social capital" debates, as an effort to alter the perceptions about the lack of participation by developing states while continuing to dictate a neoliberal approach in those states.

THE UNITED NATIONS

NGOs should not stop lobbying for major reforms of the IMF and WB, but the United Nations is their best ally in building social capital between themselves, the global justice movement, and the global governance complex. The United Nations has a culture of international assistance, equal voting rights for developing countries, and structural participation mechanisms that can lay the basis for sustainable, democratic governance. After decades of lobbying the United Nations, NGO groups have gradually attained access to the organization. At the 1972 Stockholm Conference organized by the United Nations, environmental NGOs were for the first time recognized by nation-states as legitimate actors on the global stage. Over the years, they have deepened ties with the U.N. Conference on Environment and Development (Conca, 1996; Gençkaya, 2000, p. 74). A watershed for inroads into global governance was the 1992 Rio Earth Summit, where at least twice as many (20,000) activists participated and placed transparency at the core of their demands.

Given the continued commitment to deregulation in the IMF, WB, and WTO (albeit tempered by a PWC), NGOs might be able to articulate a democratic and environmental vision of global governance at the United Nations if institutional reforms along the lines suggested by a recent United Nations University study were adopted (Nayyar, 2000). Stiglitz and others in the study call for a significant reorientation of the practices and ideologies in the Bretton Woods institutions, including greater transparency and public participation, more equal voting rights for developing countries in the IMF and World Bank, and civil society participation through a global parliament under U.N. auspices. An Economic Security Council and a Global Peoples Assembly would run parallel to the U.N. General Assembly and take on tasks now performed by the WB and IMF. Down the line, perhaps a reinvigorated United Nations could oversee the Bretton Woods institutions.

CONCLUSION: GOVERNANCE AND SOVEREIGNTY

This chapter has sketched several cases from the global institutional contexts in which NGO influence efforts have been made. NGOs and other civil society organizations appear to be more influential than ever, but their capacities are constrained by the traditional functions of intergovernmental institutions and treaties in the international economy that remain essentially unchanged. Although roundly criticized internally and externally, IMF prescriptions remain intact, as does the elite consensus around free trade. Market pieties in these contexts have not been fundamentally rethought. As the PWC shakes out in the Bretton Woods organizations, fundamental changes such as total debt cancellation or equalized voting procedures are off the table. Too often, civil society is not seen as a deep seat of representation and accountability, but rather as a location or effect of market imperfections—a type of social capital that can be managed to make governments and intergovernmental institutions efficient and effective (Phillips & Higgott, 1999).

Scholars studying forms of public participation in global environmental and economic governance institutions may find value in the notion of social capital as it relates to expanding networks of trust within and between the array of multiissue transnational NGO coalitions, the popular backlash, politicians, corporate leaders, and international institutions. The fusion of cultures through the Internet has given NGO networks the ability to challenge the policies and programs of authoritative institutions. By drawing on and fostering public demands for transparency, NGOs have rendered closed-door negotiations led by corporate elites in global governance institutions increasingly controversial. But scholars should be careful not to overemphasize the value of social capital as a policy tool. Its use in this regard seems to have become fully colonized (within the WB anyway) by neoclassical economists who are not likely to revise WB practices in favor of disempowered groups.

Nor will a celebration of social-capital building between NGO networks and the major interstate institutions resolve the uneasy tension between traditional state authority and new forms of democratic legitimacy. In a context in which flows of commerce and information have already strained sovereignty, NGO authority rests on a claim to represent cross-national interests such as environmental health, sustainable development, and human rights. The growth in transnational alliances has enhanced their claim to representation, but NGOs can still only approximate the legitimacy claimed by national states, especially in cases in which NGOs are less democratically organized and accountable than their participatory rhetoric suggests (Gordenker & Weiss, 1995; Simmons, 1998). For NGOs to assume the responsibility that comes with authority and legitimacy, the principle of sovereignty may have to be adapted to multilateral interdependence (Commission on Global Governance, 1995, pp. 68–72).

NOTES

1. I am referring to conditioning development and balance-of-payments loans to reductions in basic social services (food, health, education), privatization of state companies, reducing taxes and tariffs on imports, export-led development (exploiting cheap labor and natural resources versus subsistence or sustainable economies), currency devaluation, and limiting labor power.

2. Cf. Francis Fukuyama's "The End of History" (1989), Thomas Friedman's *The Lexus and the Olive Tree* (1999), and Mort Zuckerman's *A Second American Century* (1998).

3. Martin Khor of the Third World Network obtained a document from the May 1995 OECD Ministerial Meeting indicating that multilateral investment negotiations might proceed in the OECD. Khor passed this information out to various colleagues, including Tony Clarke of Canada's Polaris Institute in

Canada. Clarke got a draft of the MAI, translated it into a readable document, and posted it on an international e-mail distribution list, along with his own analysis and interpretation. Lori Wallach of Public Citizen's Global Trade Watch engaged in similar interpretive efforts (Tieleman 2000, pp. 12–13).

4. This focus may have a greater impact in the long term because it alters the context in which project managers act by allowing opponents to challenge and influence projects early in the design stage and leverage public opinion against the WB's failures to meet their own policies (Brown & Fox, 2000, p. 6).

5. Fine (2000, 2001) explains that the notion of social capital derives inescapably from rational choice theories, and should be understood in the context of the colonization of the social sciences by neoclassical economics. This mainstream economic theory, a reaction to the stagflation crisis of the 1970s, is based on the idea that optimizing and efficient actors counteract state intervention by rationally anticipating its impacts. This rational-expectations model suggests that all human activity, including nonmarket, social, and even irrational behavior, can be understood on the basis of methodological individualism, thus widening the analytic scope of economics to include the other social sciences (2001, p. 10). Although social capital "purports to civilise economists by forcing them to take account of the social . . . it opens the way for economists to colonise the other social sciences by appropriating the social in . . . ways with which it is entirely comfortable" (2000, p. 199). This process has been facilitated by the social sciences' retreat from postmodernism and return to material reality (2001, pp. 12–15; 2000, p. 7).

Chapter 12

Toxic Tours: Communicating the "Presence" of Chemical Contamination

Phaedra C. Pezzullo

> Before the judicial body here makes a decision, we strongly urge that you come to our city and meet with us and see where we live and see what we're exposed to. Right now, I'm offering that invitation. I would very much like an answer.
>
> —Zulene Mayfield, Chester Residents Concerned for Quality Living

> A king sees an ox on its way to sacrifice. He is moved to pity for it and orders that a sheep be used in its place. He confesses he did so because he could see the ox, but not the sheep.
>
> —Meng-Tseu

Throughout the U.S. environmental movement, various rhetorical strategies have been chosen to attempt to explain the worth of a place—particularly when such places have appeared threatened by human activities. In addition to writings, speeches, paintings, photographs, and protests,[1] tours have provided a compelling medium of persuasion for environmentalists. Since at least the early 1900s, environmentalists have invited people to travel with them on guided journeys to learn more about a particular place.[2] The primary logic behind using tours as an advocacy strategy is reminiscent of Meng-Tseu's story of the ox and the sheep: If a person exposes her or his senses to a place (or any other potential "sacrifice"), that person will better appreciate its value and, thus, will feel more connected to its fate.[3]

Although employing tours as an environmental advocacy strategy began as a means to protect more traditional environments such as wetlands and forests, the scope of tour topics has expanded as the environmental movement itself has broadened its reach. With the rise of grassroots environmental activism against toxic pollution since the late 1970s, for example, activists and scholars have argued that people of color and low-income communities have disproportionately carried the burden of waste produced in the United States and globally (Pezzullo, 2001).[4] One increasingly common mode of communicating this claim is a variation of previous environmental advocacy tours called *toxic tours*.[5]

Toxic tours are noncommercial expeditions organized and facilitated by people who reside in areas that are polluted by toxins, places that Bullard (1993) has named "human sacrifice zones" (p. 12).[6] Residents of these areas guide outsiders, or tourists, through where they live, work, and play to witness their struggle. Like other environmental advocacy tours, therefore, toxic tours provide an occasion for community members to persuade people (who they believe either directly or indirectly have the power to alter their circumstances) to better appreciate the value and, thus, the fate of their environment. For instance, Zulene Mayfield's invitation to the state Environmental Hearing Board was an attempt to bring them to her community before they passed judgment.[7] As strategic efforts to influence public judgments regarding the environment, toxic tours appear pertinent to the discussion of environmental decision making more generally.

Steelman and Ascher (1997) categorize four broad types of environmental decision making: (1) standard representative policy making; (2) referenda (e.g., initiative, referendum); (3) nonbinding direct involvement (e.g., public comment periods, hearings, some citizen advisory commissions); and (4) binding, direct policymaking by nongovernmental representatives (overseen by elected or appointed officials) (p. 71). A fifth, often unmentioned, category relevant to the public participation literature is nonbinding indirect involvement, or what most of us refer to as *advocacy*. Although not initiated by the government, environmental advocacy is certainly a means of holding those in power, including the government, more accountable to public input. My assumption, therefore, is that toxic tours, as a form of environmental advocacy, are relevant to discussions regarding public participation in environmental decision making. Furthermore, I contend that toxic tours are complex rhetorical strategies through which advocates provide a powerful critique against dominant discourses of toxic waste and the excesses of modernity more generally.

Antitoxic advocacy faces particular constraints that vary from other forms of public participation. Thus, I begin by exploring the historical context of toxic pollution relative to the issue of rhetorical constraints. Then, I briefly review tourism literature to investigate how tourism might provide not only a means for addressing the practice of touring itself, but also a metaphor for modern life more broadly. Third, I provide an analysis of Perelman and Olbrechts-Tyteca's (1958/1969) rhetorical theory of "presence" to illustrate how this concept provides a useful spatial discourse for understanding the ways in which particular arguments are intended to work. Here, I argue that the rhetorical concept of "presence" provides a heuristic useful for understanding the stakes of toxic tours. Fourth, having established a theoretical and historical context for my study, I elaborate on my earlier definition of toxic tours by identifying and describing the breadth of this practice. In closing, I consider what conclusions may be drawn from this initial study and possibilities for further research.

THE BAGGAGE OF TOXICS

Toxics, for many, are beyond human comprehension. Frequently, citations of the mere scale can overwhelm us:

> The latest data from U.S. EPA indicate that manufacturing firms released over 7 billion pounds of toxic chemicals to the environment in 1998. (Environmental Defense, 2000)

> There are some 70,000 chemical substances in commercial use, with about 1,500 new chemicals introduced annually, most of which have come into use since the end of World War II. (Field, 1998, p. 92)

> The Conservation Foundation reports that every year the United States generates approximately 50,000 pounds of air, water, and solid wastes *per each of the existing 240 million U.S. residents*. . . . Spills and other releases of hazardous materials occur in U.S. communities almost 500 times per week, or 25,000 times per year. (Bryant, 1995, pp. 46, 47; emphasis added)

In addition to the magnitude of toxic production, toxics themselves challenge our ability to perceive these substances. Reich (1991) argues that any attempt to construct a counterdiscourse to toxic production faces constraints such as invisibility of the toxic agent, nonspecificity of toxic symptoms, and difficulties of identification (pp. 142–147). Toxics sometimes are also latent in their effects. Although they accumulate in our bodies, the signs of their impact may remain dormant in the short term and become apparent only in the long term.

The one thing of which most of us are certain is that toxics are poisonous. Toxic pollution has been linked to cancer (e.g., lung, ovarian, testicular, breast, brain, and stomach), lung and respiratory disorders, central nervous system anomalies (e.g., mental retardation, cerebral palsy, and spina bifida), spontaneous abortions, reproductive disorders, depression, heart defects, asthma, skin disorders, immune system suppression, and so on (Brown & Mikkelsen, 1990/1997; Di Chiro, 1996; Grossman, 1994; Novotny, 1998; O'Brien, 2000).

Due to the excessive scale, obscurity, and danger posed by toxics, they often appear surreal for those who do not live, work, or play in or near them. Toxics, therefore, have provided suggestive fodder for fantasy, particularly in popular media. For example, the Teenage Mutant Ninja Turtles were four "ordinary" turtles who were mutated by a toxic slime that had fallen off the back of a truck and, hence, became superheroes. Their toxic encounter, in other words, enabled them to use their mutant characteristics toward good, against evil. This narrative actually is not that rare; many superheroes are the products of toxic mutation (e.g., X-Men, as products of atomic radiation; Spider-man, as a product of an irradiated spider that bit him; the Joker, as an everyday bad guy who became a super villain once he fell into a vat of toxic chemicals; etc.). Toxic chemicals, in these fantasy texts, have a two-fold nature. First, toxics serve as the catalyst for moving the world beyond our current limitations so that we may imagine flying, predicting the future ("spidersense"), feeling super-strong, gaining telepathic and telekinetic abilities, and so on. Conversely, toxics in the comics world often symbolize humanity out of control and, therefore, the superheroes constantly are fighting the forces that have mutated them to save the rest of the world from the more commonly understood effect of toxics: mass destruction.

Recently, this narrative of a larger-than-life superhero swooping into a toxic area (despite all potential personal costs) has reverberated in the Hollywood tales of *A Civil Action* (starring John Travolta) and *Erin Brokovich* (starring Julia Roberts). Although based on "true life," of all the stories regarding toxics, it is not surprising to discover which ones have captured the U.S. imagination—and how. These are not the countless stories of grass-roots communities working together to fight the industries killing them, but they are the stories focused on spectacular outsiders coming in and trying to save the locals from threatening toxic industries.

Far from fantasy, Field (1998) reminds us: "Whether it's the risk from breathing polluted air or from the consumption of contaminated fish, abstract risk manifests itself in real harm to real persons in particular places" (p. 81). For those exposed to toxics, the privilege–fantasy of escape or being "saved" no longer exists. Identifying, tracing, and naming toxic chemicals is a survival skill.

"Paradoxically," Reich (1991) suggests, "crisis in society creates a window on normality, one that offers a view of underlying political patterns not usually visible" (p. 11). Indeed, responses to this toxic crisis have illuminated the ideology of modernity in such a way that challenges our understanding of "progress."[8] Hofrichter (2000) calls this hegemonic context a "toxic culture" in which both our physical and symbolic reality is framed by "Western concepts about progress, development, private property, economic rationality, risk, science, and individualism . . . [an] emphasis on consumption over production and private over public gain, thereby transforming collective issues into personal ones . . . and institutionalized practices that constrain society's willingness to rethink what is desirable and possible" (pp. 1, 4). This toxic culture of modernity has had direct historical implications on the environment, such as

> The growth of the factory system, expansion of wage labor, increased use of machine production, and the rise of the industrial city. . . . Waterways became a convenient method for waste disposal; the air, a sink for smoke; land, a commodity that could be created (as with the extensive landfilling of the nineteenth century) and used with no restrictions except traditional legal notions of nuisance and trespass that mediate between conflicting rights of property owners. (Field, 1998, p. 89)

Reich (1991) reminds us, however, that not everyone has accepted this way of life as natural:

> On TV and in print in the 1970s, Monsanto announced: "Without chemicals, life itself would be impossible." The statement is true but misleading, a public relations attempt to counter the increasing

> popular awareness of another truth: that sometimes, because of chemicals, life itself becomes impossible. (p. 1)[9]

Furthermore, environmental justice activists have extensively argued and documented that the lives that have been disproportionately sacrificed are those of people of color and the poor (Bullard & Wright, 1987; Lavelle & Coyle, 1992; United Church of Christ Commission for Racial Justice, 1987; U.S. General Accounting Office, 1983). Thus, Bullard (1991) argues, "We have to understand the link—the correlation—between exploitation of the land and exploitation of people" (1991, videotape). For purposes of this study, it is particularly critical to note how environmental injustice turns on a real and imagined geographical segregation that enables what dominant U.S. society has deemed "appropriately polluted spaces" (Higgins, 1994). In this political pattern, those people culturally conceptualized by "white, elite centers of power" as Other are articulated with or linked to waste (which is also considered undesirable, unnecessary, and contaminating). Thus, "separate areas of existence" (Douglas, 1968, p. 40) are created, areas away from elite White existence. The invisibility of this pattern is dangerous socially and environmentally because it allows for a space to be created for nonrecognition—where people appear "file-awayable" (Lugones, 1994, p. 637) and waste forgotten. As such, accountability for what occurs in these "appropriately polluted spaces" becomes suspended.

In light of the excesses of modernity, including this pattern of social and environmental segregation, communities that have been polluted face immense challenges to being heard and recognized within dominant U.S. culture. By bringing those who live elsewhere into their community, activists are attempting to narrow the distance, literally and figuratively, between those who live, work, and play elsewhere and those who live, work, and play within these contaminated communities. Consider, for example, the reactions shared in this newspaper report of a toxic tour:

> "This part of Baltimore is not perceived by many people who live in Baltimore—it's a section of the city that isn't part of the city," said Maeve Hitzenbuhler (a school teacher and an advocate), as the chartered bus rambled past one industrial plant after another. "The community in Baltimore doesn't have a good idea of what's going on here." (Willis, 1998, p. 3B)

Similarly, another guide said:

> "The area, despite being only several miles from downtown Baltimore, is very isolated and most people on the tour have no idea of the scale of the operations, or the environmental and public health threats involved. A common reaction of tour-goers is 'I had no idea'." ("Mapping what's happening on the ground: Three toxic tour surveys," 2001, p. 10)

It is within this context of living "worlds apart" that grassroots communities engaged in toxic struggles have turned to tours.

THE PLEASURE OF TOURS

From visiting a local historical monument to visiting the Sistine Chapel, from experiencing one's first hike outdoors to experiencing one's first safari, from inspecting a potential home to inspecting a work of art, touring pervades everyday life in its most mundane and most spectacular moments. Given the range of tourism experiences, it is not surprising that for every book or article written on tourism there is a different definition of the term. In this study, I assume a *tour* is constituted by the following: physically travelling somewhere; a return to one's starting place; an experience of something different from one's everyday life; a structured interpretation of that experience; and a desire based on the assumption that the experience will be—even if only momentarily—transformative.[10]

In his foundational study, *The Tourist*, MacCannell (1976/1999) claims *tourist* is useful both as a designation for actual tourists and as a model to consider modern subjectivity more generally (p. 1). Neumann (1999) similarly argues that tourism is "a powerful metaphor for the broader character and conflicts of modern life. Tourism," he insists, "is a way of moving through the world" (p. 8). Cautioning us against universalizing a middle-class, Euro-American perspective of tourism, Kaplan (1996) reminds us, however: "Tourism must not be separated from its colonial legacy, just as any mode of displacement should not be dehistoricized or romanticized" (p. 63). Moreover, as Jordan (1982/1985, 1993) writes, "race and class and gender remain as real as the weather. But what they must mean about the contact between two individuals is less obvious and, like the weather, not predictable" (p. 312). Tourism, therefore, like modernity, suggests a complex set of relations in which various people interact with each other to persuade, silence, engage, coerce, and so on.

Much of the literature on tourism (both as a practice and as a metaphor) addresses commercial tourism, which I am not going to discuss here.[11] I am interested in the tours MacCannell (1976/1999) briefly addresses in the following passage:

> The tours of Appalachian communities and northern inner-city cores taken by politicians provide examples of *negative sightseeing*. This kind of tour is usually conducted by a local character who has connections outside of his community. The local points out and explains and complains about the rusting auto hulks, the corn that did not come up, winos and junkies on the nod, flood damage and other features of the area to the politician who expresses his concern. . . . [E]cological awareness has given rise to some imaginative variations: bus tours of "The Ten Top Polluters in Action" were available in Philadelphia during "Earth Week" in April, 1970. (p. 40; emphasis added)

This last statement regarding bus tours is the earliest reference I have found in scholarly literature of what is now referred to as toxic tours. If MacCannell had explored this phenomenon further, however, I believe he would have attributed a name more fitting than "negative sightseeing." Yes, tours of environmental and cultural tragedies are negative in the sense that they do not seem to offer the more familiar tourist package of pleasure and leisure in terms of escape from everyday life. I would argue, however, that these tours offer the possibility of an often too-rare indulgence: a sense of agency.

Many historical tours and sites, similarly, suggest a vision of hope and, thus, a call to action. The desire for a U.S. Holocaust Memorial Museum was founded not to romanticize a reading of history that might allow us to understand what happened, but to "Never Again" let a Holocaust occur (Linenthal, 1995). Similarly, the Civil Rights Memorial in Alabama was commissioned by the Southern Poverty Law Center after their initial building was firebombed by the Klu Klux Klan and displays, among other, less-conspicuous inscriptions, ". . . UNTIL JUSTICE ROLLS DOWN LIKE WATERS AND RIGHTEOUSNESS LIKE A MIGHTY STREAM. MARTIN LUTHER KING, JR." (Blair, 1999). In other words, these spaces interpolate the tourist into the outcome of the ever-unfolding stories they have begun to present. Such an invitation to agency suggests a hope that suffering need not exist, an almost utopian wish for a better world.

On what I would prefer to call *advocacy tours*, guides similarly present a particular vision of their community to their visitors so that these "tourists" might be moved to do something to transform the tragic scene presented. Rather than a staged escape from everyday life, advocacy tours instead ask their visitors to move more closely to another experience of the everyday, in the hope that they might not only be physically, but also emotionally *moved*. For example, subsequent to a toxic tour in which I recently participated on the border of Texas and Mexico, my guide said: "When we come to the border, that's where we find each other. . . . Do not be indifferent." Another guide who has provided numerous tours of the area known as Cancer Alley, LA, writes: "One of the key things that I try to impart to the persons taking the tours is that every person can make a difference if they are willing to stand up and become an activated member of the American Democracy that we live in" ("Mapping what's happening on the ground," 2001, p. 7). Ideally, through their interrogation of the ways in which the excesses of modernity affect their everyday lives, these tours perform a critique of the political economy as it is and demand that life should be better. Thus, to borrow a phrase from Bauman (1986/1987), these tours construct a "counter-culture of modernity" where guides are not reducible to locals who complain, as MacCannell's brief passage might suggest. Instead, they are weavers of bodies and places with the desire to live and the courage to risk hoping for more.

In the only academic publication explicitly focused on toxic tours of which I'm aware, Di Chiro (2000) argues the following:

> Toxic tourism can be understood as a species of ecotourism, what I have called alternative ecotourism, even though it is not a money-making venture because of its focus on the relationship between environmental degradation and social problems and the belief that firsthand experience may result in environmental action. Both toxic tourism and ecotourism challenge the remoteness of the "tourist gaze," a kind of museumlike looking from afar, and instead seek to create the conditions for an interactive form of sightseeing. Both genres of environmental tourism aim to present an experience of natural and social environments "as they really are." (p. 296)

Di Chiro's primary argument is that toxic tours are a form of ecotourism[12] because they: (1) focus on environmental and social problems; (2) believe that firsthand experience may result in action, thus challenging the remoteness of the "tourist gaze"; and (3) claim to present an authentic experience. These similarities seem indisputable. Her study, therefore, offers a useful starting place for more fully appreciating toxic tours.

Our approaches, however, differ in at least two ways. First, Di Chiro claims that "the growing popularity of the ecotourism industry has opened up potentially interesting and innovative political spaces for environmental justice organizing" (p. 279). Conversely, I locate toxic tours as part of a longer environmental movement history of advocacy tourism, rather than an outgrowth of a more recent boom in a commercial industry. Second, Di Chiro argues that the "aim is to take action to change what the eyes witness" (p. 277). I, however, am skeptical of an ocularcentric assessment of experience in general, including toxic tours. Instead, I propose that it is critical to understand the primary function of toxic tours as an embodied *communication* strategy. To elaborate, I will turn now to rhetorical theory and then back to toxic tours.

THE ELEMENT OF "PRESENCE"

Toxic tours are complex rhetorical endeavors. They include multiple rhetors. They are constituted by a variety of mediums and textures—from people's bodies to polluted waterways to odors and more. Moreover, although those organizing the tours have chosen to present certain sites and people in specific ways for particular ends, the tours are also traversed by bodies and through landscapes that are not necessarily orchestrated for the means of that tour. Given these complexities, rhetoric has much to offer one's understanding of toxic tours. One of the most useful connections between these tours and rhetorical theory, I would argue, is the concept of "presence."

In an article about former Vice President Gore's book, *Earth in the Balance*, Murphy (1994) notes that although "presence" is a tantalizing term for rhetoricians, it remains "ill-defined" and "ambiguous" both in Perelman and Olbrechts-Tyteca's (1958/1969) initial mapping of the concept and in the "scant" secondary literature (Murphy, 1994, pp. 1–2). I want to revisit Perelman and Olbrechts-Tyteca's (1958/1969) text, *The New Rhetoric*, to reestablish what ground they have covered and why it is so provocative. I argue that understanding "presence" as a spatial element provides a heuristic whereby to theorize more fully the primary function of toxic tours.

Perelman and Olbrechts-Tyteca (1958/1969) are interested in the technical aspect of presence, based on the assumption that argumentation is selective and, therefore, "cannot avoid being open to accusations of incompleteness and hence of partiality and tendentiousness" (p. 119). Thus, they continue, "[w]e must add that in the social as well as in the natural sciences this choice is not mere selection, but also involves construction and interpretation. All argumentation presupposes thus a choice consisting not only of the selection of elements to be used, but also of the technique for their presentation" (p. 120). Presence is important to argumentation because it helps the rhetor achieve her or his purpose, persuasion. This emphasis on "the relationship between audience and message," according to Leroux (1992), is what makes presence an "important concept" (p. 33).

I believe that Perelman and Olbrechts-Tyteca's (1958/1969) definition of presence also suggests a spatial understanding of argumentation.[13] "[A]ny argument, by its presence," they claim, "draws the attention of the audience to certain facts and makes it give consideration to matter that it may not have previously thought about" (p. 481). Similarly, they maintain: "The thing that is present to the consciousness assumes thus an importance that the theory and practice of argumentation must take into consideration" (p. 117). Presence, therefore, describes when an argument becomes relevant or meaningful to its audience and is important to rhetoricians because it is a vital part of the "stuff" that constitutes persuasion. "[T]he importance that should be attributed in argumentation to the role of presence," according to Perelman and Olbrechts-Tyteca, is "to the displaying of certain elements on which the speaker wishes to center attention in order that they may occupy the foreground of the hearer's consciousness" (p. 142).[14] In other words, presence occurs when an argument persuades us to feel that something matters, as if it were standing right before us, no longer able to be ignored. Presence makes that which was once seemingly absent move closer.

"Accordingly," Perelman and Olbrechts-Tyteca claim, "one of the preoccupations of a speaker is to make present, *by verbal magic alone*, what is actually absent but what he [/she] considers important to his [/her] argument or, by making them more present, to enhance the value of some of the elements of which one has actually been made conscious" (p. 117, emphasis added). A solely discursive construction of presence, as they suggest, is limiting. I, however, believe that their point is not to marginalize the nondiscursive, but to distinguish between the element of presence and reality. They continue:

> Certain masters of rhetoric, with a liking for quick results, advocate the use of concrete objects in order to move an audience: Caesar's bloody tunic which Antony waves in front of the Roman populace, or

> the children of the accused brought before his [or her] judges in order to arouse their pity. The real thing is expected to induce an adherence that its mere description would be unable to secure; it is a precious aid, provided argumentation utilizes it to advantage. The real can indeed exhibit unfavorable features from which it may be difficult to distract the viewer's attention; the concrete object might also turn his [/her] attention in a direction leading away from what is of importance to the speaker. Presence, and efforts to increase the feeling of presence, must hence not be confused with fidelity to reality. (pp. 117–118)

Here, Perelman and Olbrechts-Tyteca remind us that presence should not be confused with reality because the persuasive appeal of any given object when placed in front of an audience is not inevitable, whereas presence, by definition, is the effect felt when an audience has been persuaded that an argument matters. In other words, although a material object may help a rhetor to invoke a feeling of presence for her or his audience, it need not necessarily do so.

Disarticulated from reality, Perelman and Olbrechts-Tyteca note the importance of symbols to the element of presence:

> The symbol is generally more concrete, more manageable, than the thing symbolized. This makes it possible to exhibit in concentrated form toward the symbol an attitude toward the thing symbolized which would require long explanations in order to be understood. The act of saluting the flag is an illustration. The technique of the scapegoat simplifies behavior by making use of the symbolic relation of participation between individual and group. Not only is the symbol easier to handle, it can impose itself with a presence that the thing symbolized cannot have: the flag which is seen or described can wave, flap in the wind, and unfurl. (p. 334)

A symbol, in this sense, appears as a synecdoche for abstract ideas such as a nation. The symbolic, they contend, is one useful strategy among many to note when preparing or studying an argument. I will return to symbols subsequently.[15]

For now, it is worth noting that, in light of this reading, the element of presence offers several insights into more fully appreciating toxic tours. For example, toxic tours are selective and thus "open to accusations of incompleteness and hence of partiality and tendentiousness" (Perelman & Olbrechts-Tyteca, 1958/1969, p. 119). It is no surprise that the governor of

Ohio's communications director responded to a toxic tour with the following criticism: "A silly stunt and posturing will do nothing to address Ohio's environmental challenges. . . . We don't need them to conduct a tour for us or anyone else" ("Group blasts," 1992, p. 8B). This explicit denial of the selective nature of anyone's reality is predictable when one understands toxic tours through a rhetorical perspective. Not to be confounded with merely mirroring reality (*mimesis*), therefore, toxic tours are interpretive constructions of a particular understanding of environmentally related practices and effects (*poeisis*). Although one of the primary facets of this strategy, of course, is going there, Perelman and Olbrechts-Tyteca (1958/1969) remind us that concrete objects alone do not guarantee favorable results. It is the ways in which tourists are guided through these spaces physically and orally that make toxic tours matter.

THE PURPOSE OF TOXIC TOURS

Depending on the rhetorical situation, toxic tours are organized by a variety of groups for an assortment of reasons. Environmental-justice[16] grassroots groups have provided tours for and with a broad range of audiences such as academics, mainstream environmental groups, political representatives, government officials, and media reporters. Typically, they choose one or two guides to walk or drive from block to block, pointing out where polluting industries are located in relation to the residents, stopping to allow the tourists to witness the stories of various residents' ailments and struggles, and providing information they have gathered regarding the violations and the apparent effects of these industries on the surrounding land and people.

At a 1993 press conference that occurred in conjunction with a toxic tour of her community, Charlotte Keyes, a leader of the grassroots group Jesus People Against Pollution (JPAP), stated:

> Many of us who are suffering from skin rashes, cancer, kidney failure, lung disease—you name the sickness, and these communities who have been toxic poisoned, we have it. And yet, they tell us that we have to prove that the sickness is related to the site. Now this is what I call a terrible situation. Why do we have to *prove* anything when you perfectly well know what you have done? You have performed genocide on all of the poor people of America. And I say this in His Grace, before God. And I say today, standing here before all of you, . . . that justice will prevail. (1995, videotape)

Drawing on the cultural resonance of religion and, thus, morality, Keyes's speech protested both the burden of waste and danger and the burden of proof placed on her community and provided a rhetorical vision of justice. Along with this speech, JPAP documented health effects through filming a tour of their community and collecting over 20,000 environmental health surveys (videotape). Hence, in addition to drawing media attention, the tour afforded an occasion to provide evidence of the harm toxics have caused in their everyday lives.

Toxic tours go beyond what people can see and hear, however. For example, one of the most common observations during these tours is odor. On a tour of Chester, PA, a child who lived in the area remarked: "If you want to ask my opinion, I say it pretty much stinks 'cause I want this place to go down. . . . I'm outside playing all day and all of a sudden it starts stinking and I've gotta go inside. [He pauses again and shakes his head.] Where should I go?" (Bahar & McCollough, 1996, videotape). A Sierra Club guide in Memphis notes that toxic tours provide "firsthand" evidence of "the environmental insult to residents (of having these polluters so close to homes), as well as the noxious odors that permeate the neighborhood" ("Mapping what's happening on the ground," 2001, p. 8). These odorous fumes often cause residents' and their visitors' eyes to water and throats to tighten (Setterberg and Shavelson, 1993), a reminder of the physical risk toxics pose.

Sometimes, local communities and national organizations collaborate to pool resources and energy together to present a tour. When asked to meet with the governor of Louisiana, for example, Greenpeace organizer Damu Smith (1999) responded: "Greenpeace would indeed like to meet with you. However we prefer to meet with you along with representatives of the many people in Louisiana whose lives, families and communities are being severely harmed by your environmental policies. We hope you are willing to accommodate our request" (p. 1).[17] The need for the governor to meet with community members in addition to Greenpeace was subsequently highlighted in a Greenpeace Toxics Patrol, a seven-day bus tour that began with a press conference in the state capitol and continued to travel to polluting facilities such as a napalm incinerator (Rhodia) and two vinyl producers (Formosa and Dow) (Greenpeace, 1999a). The tour, thus, provided an opportunity for those who live(d) outside of affected areas to move closer physically and conceptually to the toxic harm caused in the everyday lives of numerous Louisiana residents. In fact sheets distributed during the tour and posted on their Web site, Greenpeace (1999a) drew on the Toxic Release Inventory database to illustrate the scale of pollution in the state:

"Among the U.S. states [*sic*], Louisiana ranks first in per-capita toxic releases to the environment—first in water releases, second in total chemical releases, third in air releases, and second in amount of wastes injected into the ground" (p. 1). In addition to statistics, tour participants witnessed testimony from residents:

> One person who attended the rally made many of us want to cry (and to work even harder). Abigail was a beautiful, sweet two-year-old with a badly deformed foot and ankle, and a leg that was shorter than the other. Her mother and grandmother are convinced the deformities are the result of toxics that invade their Mossville community from Condea Vista, PPG and Conoco. The sight of this sweet child made it hard to suffer the callous, chuckling industry reps that attended. (Greenpeace 1999b, p. 1)

This juxtaposition of who benefits from, and who pays for, toxic industries is one of the powerful potentials of a tour.

Guides also frequently move between the concrete and the abstract by, on the one hand, drawing on the power of naming while, on the other hand, also allowing certain images to speak for themselves. One account of a tour organized by the Sierra Club relays the following snapshot of a schoolteacher who had to take a leave of absence due to her health: "'I am a victim of this,' she said, motioning her hand at the chemical plant" (Swerczek, 1999, p. B3). This resident simultaneously claimed her status as a "victim" (which suggests wrongdoing and the need for redress) and, with a rhetorical gesture of her arm, allowed the chemical plant to symbolize "this" (or of what she is a victim). Thus, "this" becomes a synecdoche for all that has threatened her body and livelihood, all that looms over her entire community's health and quality of life. The chemical plant is not merely mirrored for the audience, but rhetorically framed as the perpetrator of a crime.

Johnson (1998) reports of another toxic tour that was based in Louisiana's "Cancer Alley." She focuses on a single-day event organized by the National Council of Churches of Christ for religious leaders. Its purpose was not only to move tourists closer to affected communities, but to do so by articulating environmental justice as a religious mission. After the tour, "one of the ministers mention[ed] the haunting sight of children playing in the shadow of the Shell complex, and around the table [were] murmurs of shock, outrage, sorrow. Bishop P. A. Brooks of Detroit . . . [shook] his head. 'What this is about,' he says, 'is children. It is about the earth. It is about life.' Children. The earth. Life. More than any other commonality," Johnson (1998) writes, "it is these three elements that bind the work of the religious and environmental communities" (p. 7). Dialogues such as this illustrate how toxic tours offer powerful forums for strengthening coalitions.

Finally, it is worthwhile to note that in addition to tours of and by nonprofit organizations, toxic tours have become incorporated increasingly in public participation gatherings facilitated by the government.[18] For example, at the Environmental Justice Enforcement and Compliance Assurance Roundtable sponsored by the Enforcement Subcommittee of the National Environmental Justice Advisory Council (NEJAC) and the U.S. Environmental Protection Agency (EPA) in San Antonio, Texas, October 17 through 19, 1996, a three-hour bus tour was organized by local community organizations on the first day for approximately 95 people. Within that time, participants met with representatives of over a half-dozen grassroots environmental groups at locations ranging from a local church to a local Superfund site. According to the EPA Report (US EPA, n.d.) the goals of the tour were:

(1) To provide representatives of the EPA's Office of Enforcement and Compliance Assurance, EPA Region 6, and the Texas Natural Resource Conservation Commission (TNRCC), among others, a glimpse of the concerns and conditions of citizens living near environmental justice sites;

(2) To educate government representatives and provide examples of environmental racism in such communities that stem from a failure to enforce environmental regulations;

(3) To allow community grassroots organizations the opportunity to share strategies for responding to environmental injustice; and

(4) To strengthen the environmental justice movement. (p. 3)

From empowering community voices to educating outsiders through a "glimpse" to fostering the environmental justice movement more broadly, these purposes provided by the EPA further illustrate the many ways in which toxic tours offer a complex strategy for public participation.

CONCLUDING POSSIBILITIES

As Gaventa (1999) reminds us, public participation "does not just happen, even when the political space and opportunities emerge for it do so.

Developing effective citizenship and building democratic organizations take effort, skill, and attention" (p. 50). Although the excesses of modernity and touring as a practice in general appear prolific in these times, activists who construct, organize, and facilitate toxic tours have begun to articulate a powerful mode of rhetorical invention that should not be taken for granted.

Given the prevalence of this strategy and the lack of research focusing on it, this chapter offers more possibilities for further research than it does conclusions. First, toxic tours challenge the business-as-usual attitude by those who support toxin-producing industries. By bringing worlds together, these tours collapse the false separations between production and waste, wealth and poverty, and privilege and race, to illustrate how these spheres are dependent on and, thus, obligated to each other. No longer merely overwhelming and imperceptible, toxics are rearticulated as "real" and something that is worth tracking. For further research, one might ask what cultural resources are drawn on in these various communities. In other words, what is the relationship between the local and this more general practice? In addition, while it appears worthwhile to pursue mapping—what it is precisely that could be done or changed according to the vision of those who conduct toxic tours? What laws about toxics are challenged? What cultural practices? What values?

Second, as MacCannell (1976/1999) argues, tours provide a means of considering modernity more broadly. In the case of tours focused on toxics, the attention is turned to the pattern of excess that has become prevalent in modern times. Waste, human suffering, and environmental degradation are all increasing costs of modern industrialization. By constructing tours of the prices that are being paid, environmental justice advocates are rhetorically inventing a counterculture to nothing less than modernity itself. One further area of research beyond the scope of this initial study would be to map more specifically the tensions and modernist tropes that these tours are attempting to address (e.g., invisibility of toxic agents and corporeal evidence; scientific reason and human experience, etc.).

Third, because the pattern of toxic domination is predicated on separate spheres of existence, I have established that the primary function of toxic tours as a mode of rhetorical invention is to communicate a sense of what Perelman and Olbrechts-Tyteca (1958/1969) call the element of "presence" in argumentation. It, thus, is not merely enough to see the costs of toxic industries, but rather guides must articulate images, smells, feelings, and so on in such a way that when the visitor leaves (as all do, by definition), the articulation still matters and is not left behind. Because this study emphasized the purpose of toxic tours, future research might focus upon the effects: which moments move participants to sense "presence"? Does this matter when concrete environmental decisions are made? If not, why do people continue to practice this strategy?

In conclusion to his study on responses to chemical disasters, Reich (1991) provocatively states:

> Disaster may provide the opportunity and the impetus for institutional change, allowing for a redistribution of power and a transformation of policy. But the existing distribution of power creates formidable blockages to social change that might benefit the relatively powerless. To obtain even incremental changes in laws, institutions, and goals, more social crisis may be required. To prevent more chemical disasters, paradoxically and tragically, we may need more chemical disasters. (p. 281)

Given this study and the possibilities it suggests for further research, I would like to propose a less depressing alternative. To prevent further disasters, perhaps we need to make the communities that bear the costs of toxic pollution more present to those who live, work, and play elsewhere. Perhaps we need to move closer.

NOTES

First epigraph: Zulene Mayfield, Chester Residents Concerned for Quality Living, in closing her remarks to the Environmental Hearing Board in Harrisburg, PA, 1996.

Second epigraph: Summary of a story by Meng-Tseu, quoted in Perelman and Olbrechts-Tyteca, 1958/1969, p. 116.

1. For environmental communication scholarship on these various forms of advocacy, see writings (Cooper, 1996; Killingsworth & Palmer, 1996; Opie & Elliot, 1996; Oravec, 1996; Sandmann, 1996; Slovic, 1996; Ulman, 1996), paintings (Clark, Halloran, & Woodford, 1996; Oravec, 1996), photographs (DeLuca & Demo, 2000; Neumann, 1999), protests (DeLuca, 1999; Short, 1991), and other forms (Kaminstein, 1996; Kraft & Wuertz, 1996; Peterson, 1998; Peterson & Peterson, 1996; Weaver, 1996).

2. One of the most famous examples was captured by U.S. President Theodore Roosevelt (1913/1920) in his autobiographical account of John Muir guiding him on a three-day trip in Yosemite. As Nuttal (1997) reminds us, Muir himself was an avid traveler and tourist (p. 225). Stephen T. Mather, the first director of the U.S. Park Service, additionally wrote of the advocacy worth of going to U.S. parks: "our parks are not only showing places and vacation lands

but also vast schoolrooms of Americanism where people are studying, enjoying, and learning to love more deeply this land in which they live" (quoted by Olwig, 1996).

3. Meng-Tseu's story is offered as an anecdote. Tours for the purpose of advocacy vary from this story in many ways; for example, they tend to highlight an existing or potential sacrifice and, instead of offering another sacrifice, they often argue no other person or place should have to bear such a burden.

4. Although this pattern is global, I am focusing on the United States for this study.

5. I am unsure of the origin of toxic tours. Newspapers provide evidence of naming this practice as such since the late 1980s (Spears, 1989). Although not necessarily called "toxic tours," tours have been used as a form of antipollution advocacy since at least the late 1960s (*New York Times*, 1969a, 1969b, 1969c).

6. Bullard (1993) claims such areas share two characteristics: "(1) They already have more than their share of environmental problems and polluting industries, and (2) They are still attracting new polluters" (p. 12).

7. Then Chairman of the Board, Rep. Jeffrey Piccola, accepted Mayfield's offer (Bahar & McCollough, 1996).

8. For an elaboration on "progress" as an ideograph, see DeLuca, 1999, pp. 46–51.

9. Modern abuse of the environment has been masked increasingly and consistently by media campaigns assuring that the public interest was and is of primary concern to these rapidly developing industries (*Journal of Advertising*, 1995; Goldman & Papson, 1996).

10. By focusing on tourism and not "travel" more broadly, I temporarily am excluding a discussion of the much larger spectrum of travel, including commuters, migrants, and so on.

11. For a concise list of the vast range of environmental commercial tour experiences, see Davis, 1997, p. 11.

12. "As defined by the Ecotourism Society (established in 1991), ecotourism is 'responsible travel that conserves natural environments and sustains the well-being of local people'" (Di Chiro, 2000, p. 278).

13. Murphy (1994) similarly argues, "presence possesses a kind of magical quality, one difficult to describe in discursive academic language and one that is, perhaps, best represented by the implicit metaphor in its name. An auditor 'feels' the argument; it almost seems to be in the room" (p. 5).

14. In a related observation, Burke (1966) had previously noted what is now a well-rehearsed rhetorical adage: "Even if any given terminology is a *reflection* of reality, by its very nature as a terminology, it must be a *selection* of reality; and to this extent it must function also as a *deflection* of reality" (p. 145).

15. Perelman and Olbrechts-Tyteca (1958/1969) elaborate on the relationship between presence and various other choices of argumentation. Rather than

listing all facets, Murphy (1994) argues: "To try to theorize presence is to drain the vitality from the concept. Each speech will create its own sense of presence differently (if at all!). . . . An understanding of presence requires criticism, not definition" (p. 13).

16. I choose the name "environmental justice" rather than "antitoxic" because the tours I have chosen to focus on identify with that branch of the environmental movement more than any other. In addition, with the name change from CCHW (Citizens' Clearinghouse for Hazardous Waste) to CHEJ (Center for Health and Environmental Justice), it appears the primary national organization of the antitoxics movement is also privileging the rhetoric of environmental justice.

17. Smith's statement performs an ethic of "speaking for others" that is critical to partnerships between mainstream environmental organizations and environmental justice communities (Alston, 1990; Di Chiro, 1998; Gottlieb, 1993).

18. As part of the Environmental Justice Small Grants Program, the EPA also has provided funds to facilitate toxic tours. See, for example, grant number: EQ-993450-01, the Anacostia Watershed Society's request for funds for "[a]n environmental justice tour of D.C. for 32 members of the Coalition on Environment and Jewish Life" (p. 15). [On-line]. Available: http://es.epa.gov/oeca/oej/success.pdf

CHAPTER 13

ART AND ADVOCACY: CITIZEN PARTICIPATION THROUGH CULTURAL ACTIVISM

JOHN W. DELICATH

Traditional approaches to public participation in environmental decision making emphasize citizen involvement in institutionalized settings with specific mechanisms and forums for engagement with government officials and other stakeholders. Typically these approaches exclude protest activities and other citizen actions as outside the scope of specifically organized formats.[1] Although such conceptions of public participation have value in exploring particular mechanisms and forums for citizen involvement, they are unnecessarily narrow if one seeks a more complete account of citizen involvement in environmental decision making. More specifically, approaches that exclude protest activity and other forms of citizen advocacy leave unexplored the relationship between public participation in institutionalized contexts and the skills, knowledges, and emotions involved in citizen advocacy outside the specific forums of government. Furthermore, traditional accounts of public participation would further marginalize the cultural resources of protest and relegate to the periphery of public participation citizen involvement in the form of art, theatre, music, and dance. I believe this is a mistake.

The purpose of this chapter is to explore the role of cultural activism in the context of citizen involvement in environmental decision making. Hofrichter (1993) defines cultural activism as a "form of political activity practiced by many grassroots groups that serves as a face for change in that struggle. It represents a way of giving voice to people in their own language and images, derived from historical memory and current experience" (p. 89). Cultural activism involves the use of cultural resources (art, storytelling, theatre) as a means to broaden the range of expression in environmental controversies and to challenge assumptions about the organization of society.[2] It is my contention that our understanding of citizen involvement in environmental decision making would benefit from considering how cultural activism (and other forms of citizen protest) attempts to hold those in power (government and industry) accountable to the concerns of the public. The point is not to broaden the concept of public participation so wide that it includes every form of organizing and advocacy, but to explore the potential relationships between participation in institutionalized forums and advocacy activities in noninstitutionalized contexts. This chapter suggests that cultural activism is an important dimension of citizen involvement, not something peripheral to participation in institutionalized decision making. I argue that cultural activism can be understood as a tool for organizing, a resource for articulating community voice and vision, and a means of both creating opportunities and preparing for public involvement in institutionalized contexts.

To highlight the potential relationship between citizen advocacy, protest activities, and public participation in environmental decision making, I examine the role that a photographer and a photographic exhibit played in the struggle for environmental justice in a small Texas town. *Fruit of the Orchard* is series of photographs documenting the suspected health effects of a chemical disposal facility on the residents of Winona, Texas. The photographs were taken by Tammy Cromer-Campbell as part of her work with the grassroots environmental justice group Mothers Organized to Stop Environmental Sins (M.O.S.E.S.).[3] Seen through the lens of cultural activism, the efforts of Cromer-Campbell to create and exhibit her photos constitute a powerful form of citizen advocacy and play an important role in citizens' efforts to participate in environmental decision-making.

MOTHERS ORGANIZED TO STOP ENVIRONMENTAL SINS AND THE STRUGGLE FOR ENVIRONMENTAL JUSTICE IN WINONA, TEXAS

Winona is a small, rural community in east Texas. It has a population of approximately 500 residents, and including adjacent communities, approximately 1,800. It is located roughly 100 miles east of Dallas and 100 miles west of Shreveport, Louisiana. The community's residents are mostly poor and largely African American. Wallace Bluff, the section of Winona nearest to the facility in question, is inhabited mostly by descendants of slaves freed after the Civil War.

In 1982, Gibraltar Chemical Resources came to town. The community was originally told that Gibraltar would install a saltwater injection-well facility and plant fruit orchards on the surrounding land. Instead, trucks and trains from across the United States and Mexico rolled into Winona to dump millions of gallons of toxic material down two narrow, one-mile-deep shafts, where it was held in a supposedly impermeable rock formation. No fruit orchards were ever planted.

The Winona facility consisted of two commercial hazardous-waste injection wells, a hazardous-fuel blending operation, and a solvent-recovery facility. Since its inception, residents of Winona complained about the odors from the facility and suspected that an increasing myriad of medical problems, including asthma, cancer, and birth defects were caused by the chemicals being injected into the wells and released into the air.

Residents of Winona, mostly its children, suffer from numerous health problems: birth defects, rare tumors and cancers, stunted growth, brain and liver damage, kidney abnormalities and malfunctions, skin discolorations, immune deficiencies, chromosomal abnormalities, respiratory problems, and multiple chemical sensitivity. Many residents attribute the unexplained illnesses and conditions to operations at the facility.

The facility did have a lengthy history of air-purity and hazardous-waste violations. An Environmental Protection Agency enforcement officer once called it a "horror to its neighbors" (see Hewitt & Calkins, 1998, p. 55). On October 18, 1991, as Phyllis Glazer was driving her son to school, they passed the chemical plant. Thick, reddish-brown clouds were belching into the air and, more alarmingly, scores of workers were fleeing the plant. The fumes caused a burning sensation in Glazer's nose and throat, and within a few days she was shocked to see a large hole developing in her nasal septum. After the same emission that spooked Glazer, an older woman began lactating as if she were pregnant and her chickens stopped laying eggs.

Between 1982 and 1992, the Texas Natural Resources Conservation Commission received over 100 complaints about the facility. In November 1992, the Texas Attorney General's office filed a lawsuit demanding that it comply with the Texas Clean Air Act. The company eventually agreed to pay $1.1 million in fines. Then, in January 1993, an accident at the plant sent a huge cloud of corrosive hydrogen bromide and other potentially dangerous chemicals into the air. In 1994, the facility was purchased by

American Ecology Environmental Services Corporation. In March 1997, American Ecology officials announced they would close operations at the Winona facility and cited opposition by M.O.S.E.S. as the reason.

Almost since its inception, residents in Winona complained about the facility. For years residents voiced their concerns about the odors emanating from the plant and about the potential dangers associated with the chemicals being injected into the ground at the facility. Previous attempts to organize and address environmental health concerns associated with the operations at the facility produced few tangible results.

In 1992, Phyllis Glazer founded M.O.S.E.S., or Mothers Organized to Stop Environmental Sins. In what is now a familiar pattern in the movement for environmental justice, Glazer, a previously apolitical wife and mother of three, helped found an organization to formally protest the operations at Gibraltar. A few years after her family purchased a ranch in the town, Glazer began building on the environmental activism of others in Winona. She orchestrated weekly protests, and paid the fees of attorneys and experts hired to file lawsuits on behalf of the town's residents. Glazer also orchestrated bus trips to demonstrate at the state capital in Austin. On two occasions, Glazer organized bus trips to Washington, DC.

Although the American Ecology facility is closed, the controversy surrounding the effects of the plant on the community's health rages on. American Ecology has filed a multimillion-dollar lawsuit against Glazer alleging a criminal conspiracy to interfere with the operations of the business. Glazer and other residents filed a $200 million class-action lawsuit of their own charging that the facility poisoned residents through chemical releases into the air and water. Both lawsuits are still tied up in the courts. American Ecology continues to monitor the injection wells at the Winona facility. Meanwhile, residents are still getting sick and others are dying. Glazer herself is now sick, having recently been diagnosed with a brain tumor. M.O.S.E.S. continues to call attention to what happened in Winona, to lend what support they can to other communities facing dangers associated with chemical exposure, to demand that citizens play a role in decisions affecting community health, and to advance the larger cause of environmental justice.

FRUIT OF THE ORCHARD AS CULTURAL ACTIVISM

Fruit of the Orchard (FOTO) is the result of photographer Tammy Cromer-Campbell's efforts to help M.O.S.E.S. publicize the situation in Winona. She was asked by M.O.S.E.S. in 1994 to do pro bono work for the group, photographing Jeremy as a poster child for its campaign to raise public awareness and garner public support for its efforts to close the facility. Nearly a year passed before Cromer-Campbell contacted M.O.S.E.S. about photographing other residents of Winona. A concern for the residents and their well being motivated her to go back and photograph other affected residents so that she could document their story on film as well.

Fruit of the Orchard (FOTO) consists of 58 finished, black-and-white photographs. Between 30 and 35 images form the basis of the traveling exhibit. Most of the photographs in FOTO are of the families and children of Winona. The images of the Winona residents in FOTO are rich, intimate portraits of children and families who believe they are victims of chemical exposure from the operations of a hazardous-waste disposal facility. The photographs are at once both beautiful and frightening.

FOTO also includes photographs of the American Ecology facility, capturing ghostly images of the clouds billowing from the thermal oxidizer at the plant and offering a unique perspective on the proximity of the facility to the homes of residents and to the public school in Winona.

FOTO also includes photographs of the demonstrations and rallies staged by M.O.S.E.S., including their trips to Washington, DC, to meet with members of government and officials from the Environmental Protection Agency. One image, "In the Halls of Congress," paints an intimate portrait of the struggle to have one's voice heard as it documents members of M.O.S.E.S. crammed into a narrow hallway, waiting to meet Administrator Carol Browner outside her office at the EPA.

Hofrichter (1993) argues that cultural activism is "a strategy for social change and liberation" that "offers people a way to reflect on their relationship to daily life and to control their images and representations" (p. 90). Politically motivated uses of photography constitute a powerful form of cultural activism. The great photographer Gordon Parks remarked that he used his camera as a weapon against all that he saw wrong with society.[4] I believe that it is in such a tradition that *Fruit of the Orchard* operates. Recognizing the power of the image to highlight the problems of pollution and toxic waste, Cromer-Campbell sought to use her photographs as a tool in the struggle for environmental justice. Her photographs are displayed on the M.O.S.E.S. Web site and used in many of the group's publications.

Although Phyllis Glazer makes for a ready-made media icon, Cromer-Campbell's photographs have given another face to the situation in Winona and provided additional incentive for the media to cover the issue.[5] Her photographs have accompanied stories about the plight of Winona residents in such magazines as *The New Crisis* (Smith, 1997), *People* (Hewitt & Calkins, 1998), and *Texas Monthly* (Hollandsworth, 1997). The photographs have also been printed in newspapers throughout the state of Texas (see Pugh, 1997).

Images from *Fruit of the Orchard* played an important role in M.O.S.E.S.'s trip to Washington, DC, for the National Stand Up for Children rally, an event sponsored by the Children's Defense Fund. After viewing M.O.S.E.S.'s exhibit at the conference, an aide for Congressman David Bonior called to set up a meeting and to organize a special media event to highlight what happens to people in affected communities.

Cromer-Campbell's photographs have also been exhibited in conjunction with her appearances at conferences on environmental justice and public participation in environmental decision making. The photographer has made a conscious effort to exhibit her work on college and university campuses and to attend conferences on environmental issues. She sees this strategy as important for continuing to tell the story of Winona and M.O.S.E.S., to make students aware of the issues of toxic waste, pollution, and environmental justice, and to build bridges between activists, academics, and policymakers.

The images in *Fruit of the Orchard*, along with the text panels that accompany the photographs, tell the story of the residents of Winona and document the efforts of M.O.S.E.S. to participate in decisions regarding the operations of the facility. In one sense, FOTO is fundamentally a story about citizen involvement in environmental decision making. It is a story that suggests that poor and minority communities, those systematically disenfranchised throughout society, often bear the health costs associated with the disposal of hazardous waste. It documents the fact that communities face serious obstacles to being taken seriously, to have their voices heard, their opinions respected, and to battle powerful forces in environmental decision making.

FOTO also demands that industry and government be accountable for protecting public health. Just as important, FOTO reaffirms a fundamental tenet of the movement for environmental justice—that people should have a right to participate in decisions that affect their lives.

CULTURAL ACTIVISM AND/AS PUBLIC PARTICIPATION

Theorizing the role of cultural activism in the context of citizen involvement in environmental decision making does not require expanding the concept of public participation broadly to consider any and all forms of political activity. It does, however, encourage us to be sensitive to the relationships between participation in institutionalized and noninstitutionalized contexts and to be cautious of the tendency to separate decision making, politics, protest, and cultural expression into separate spheres of activity.

At a most basic level, cultural activism involves attempts to make those in power (government and/or industry) accountable to the interests of the community and therefore constitutes a means of citizen involvement. More specifically, cultural activism addresses issues fundamental to the current limitations on citizens' attempts to meaningfully participate in environmental decision making such as cultivating the desire to participate and mobilizing citizens to act; developing the confidence to participate; developing skills of self-advocacy; and discovering ways to participate that respect community voice and vision.

In the following discussion I will attempt to illustrate the possible relationships between cultural activism and citizen participation in the institutionalized contexts of environmental decision making. My arguments are both hypothetical and based on the possibilities evident in the efforts of M.O.S.E.S. in east Texas. I should say at the outset that I intend to offer few empirical claims about the influence of cultural activism on the efforts of citizens to participate in the institutionalized contexts of environmental decision making. My purpose is to highlight the potentially complex ways that cultural activism can shape, replace, and sometimes supercede citizen participation in the institutionalized forums of environmental decisions.

First, cultural activism is a strategy available when participatory opportunities are nonexistent, limited, or inadequate. There are numerous instances in the history of the movement for environmental justice in which exclusion from the decision-making process or obstacles to meaningful participation in institutional contexts were countered by citizens employing the symbolic resources of protest to create opportunities to influence decision making. Staging image events, dramatic acts of protest designed for media dissemination, has been one form of cultural activism that environmental justice groups have used to create opportunities for participation (DeLuca, 1999). One might also imagine the how the use of toxic tours described in the previous chapter can be an effective tool for opening up institutional spaces for citizen involvement (Pezzullo, this volume). In this sense, cultural activism is a strategy for creating awareness, publicizing the existence of problems, and applying political pressure that may create opportunities for citizens to participate in formal contexts for environmental decision making. Additionally, cultural activism and other forms of political pressure, especially those capable of generating publicity and media attention, encourage transparency and accountability in institutionalized decision making by making issues visible to the larger public.

Second, the creative forces of cultural activism are important resources capable of energizing citizens and motivating them to get involved. As Branda Miller argues (2000), "creative practice is essential, not frivolous; cultural work is not the icing on the cake, it is a core aspect of organizing" (p. 324). Cromer-Campbell fully believes in the power of the images in FOTO, and of photography in general, to inspire people to act. When asked if she thought her photographs could help affect social change, she replied; "If my work is successful, and people take the time to look and explore what happened to this community, then maybe someone will take a stand against pollution" (T. Cromer-Campbell, personal communication, October 17, 2001). She continued, "I think that if people see the work, read the text, and are sensitive to what happened in this tiny community, hopefully it pisses them off enough to get them to their feet as activists" (T. Cromer-Campbell, personal communication, October 17, 2001). This possibility always exists with art. Murray Edelman (1995) argues that "contrary to the usual assumption—which sees art as ancillary to the social scene, divorced from it, or at best, reflective of it—art should be recognized as a major and integral part of the transaction that engenders political behavior" (p. 2).

In addition to motivating and inspiring citizens to participate in environmental decision making, cultural activism can empower citizens, giving them a sense of confidence that their efforts to participate will be meaningful. "For those who are victimized by structures in which they feel powerless, as is so often the case for communities battling large industries and government institutions over environmental health concerns, the self-awareness and self-confidence" (Miller, 2000, p. 314) resulting from creative practices such as cultural activism can play an important role in empowering citizens' ability to participate in institutionalized settings.

Third, forms of cultural activism allow citizens to speak on their own terms, to make arguments they perhaps would not otherwise be able to make, and to articulate community voice and vision in uniquely powerful ways. There are numerous reasons (e.g., social and geographical segregation, political disenfranchisement, perceived lack of knowledge and expertise, lack of access to the mass media) why polluted communities face difficult challenges being recognized and heard. By finding novel and innovative means of telling their story and reaching out to wider public audiences, grassroots environmental justice groups are developing modes of cultural activism that allow their voices to be heard and that afford them opportunities to contribute to the public dialogue on matters of environmental injustice. As in the case of *Fruit of the Orchard*, photography gives the problems of Winona a human face; it personalizes the threats from chemical exposure. It is one thing to describe the operations of the facility and to discuss the routine releases of chemicals into the air. It is another to see clouds of gas and chemicals billowing from the facility and the homes of Winona residents in the background. It is one thing to hear that the American Ecology facility is located close to the local school. It is another to see the juxtaposition of the sign pointing to the Winona school and the facility immediately behind it as in the photograph "Winona School Sign." It is one thing to hear that residents suffer from a variety of health conditions that they attribute to the plant. It is another to see the individuals, their families, and their medical maladies captured in black and white photographs that are at once beautiful and intimate, and ghostly and frightening.

Finally, cultural activism can transform the terms of public debate and affect the opinions of decision makers and thereby influence decision outcomes. Cultural activism and other forms of citizen advocacy in noninstitutionalized contexts can introduce new ideas and arguments into the controversy. Cultural activism can help articulate, disseminate, and circulate the interests, concerns, and arguments of community members. This can take place in the context of specific decisions such as M.O.S.E.S.'s position concerning the American Ecology facility or the broader context of public discourse about environmental justice such as M.O.S.E.S.'s contribution to the demand that citizens play a greater role in environmental decision making.

It is important to note that members of M.O.S.E.S. were active participants in environmental decision making in both institutionalized and noninstitutionalized contexts. M.O.S.E.S. was committed to participating in the formal mechanisms and forums for citizen involvement. Indeed, the idea that citizens have a right to participate in decisions that affect their lives is a defining characteristic of M.O.S.E.S.'s identity. As the homepage of the M.O.S.E.S. Web site reads: "M.O.S.E.S. is about citizens rightfully in charge of their own community and future."[6] Cultural activism was a complement to, rather than a substitute for, involvement in public hearings, political decision making about facility permits, and clear-air compliance forums. Members of M.O.S.E.S. were active participants in the public hearings before the Texas Natural Resource

Conservation Commission (TNRCC) concerning operating permits for the American Ecology facility. Strategies of cultural activism became a way to participate in environmental decision making beyond the scope of decisions concerning the American Ecology facility in Winona. Through the use of photographs of the situation in Winona and the exhibition of FOTO at sites around the country, M.O.S.E.S. was able to keep telling their story and to promote education and awareness about the potential human health hazards of chemical exposure. Cultural activism was also a means to make a contribution to the larger movement for environmental justice and to assert the right of all citizens to participate meaningfully in decisions affecting their communities.

CONCLUSION

Seen through the lens of cultural activism, the creative efforts of M.O.S.E.S. do more than simply call attention to the plight of the residents of Winona. M.O.S.E.S.'s work reveals that cultural activism is an important dimension of citizen involvement, not something peripheral to traditional participation in institutionalized decision making. It is evident from the efforts of M.O.S.E.S. that cultural activism is best understood as a tool for organizing, a resource for articulating community voice and vision, and a means of creating opportunities and preparing for public involvement in institutionalized contexts.

Although this case study offers little in the way of empirical evidence about the influence of cultural activism on citizens' ability to participate in environmental decision making, it does call attention to the need to explore the relationships between public participation in formal forums and citizen advocacy in noninstitutional settings. The influence of art and other forms of cultural activism on public participation in politics will always be difficult to measure. However, Murray Edelman (1995) reminds us that "art is the fountainhead from which political discourse, beliefs about politics, and consequent actions ultimately spring. There is, of course, no simple causal connection here, because works of art are themselves part of the social milieu from which political movements emerge; but there *is* a complex causal connection" (p. 2; emphasis in original).

Deliberative models of involvement and participatory approaches to public participation that emphasize the role of active citizens equipped with the knowledge and skills to take part in public life would do well to consider the issues of what motivates, inspires, prepares, and empowers the public to participate in institutionalized settings. It is not clear that informed and empowered public participation in environmental decision making is solely dependent on opportunities for institutionalized involvement. Developing innovative mechanisms, designing novel forums, and redesigning participatory processes offer no guarantees of realizing participatory ideals or achieving meaningful public participation (Owens, 2000). Gaventa is correct to note that public participation "does not just happen, even when the political space and opportunities emerge for it to do so" (1999, p. 50). Meaningful citizen involvement in environmental matters requires motivation, trust, and confidence on the part of the public. Informed and empowered public participation in environmental decision making requires inspiration, effort, and skills that require time to cultivate and develop. At both broad and particular levels, cultural activism and the ideas expressed in art and other cultural poetics (Platt, 1997) are likely to play an important role in citizens' efforts to meaningfully participate in environmental decision making.

NOTES

The views expressed in this chapter constitute the author's personal and professional opinion, and do not represent the opinion or position of the U.S. General Accounting Office.

1. See, for example, Renn, Webler, & Wiedemann (1995).
2. Cultural activism remains an undertheorized concept. In many instances the term is used without ever being defined (Bharucha, 1998). Often the term refers simply to cultural products or practices that have political significance (Felski, 1994; Goldstein, 1994). Authors who have attempted to define and theorize the term have emphasized the intentional use of cultural resources in the context of social movement mobilization (Field, 1995). The most sophisticated applications of the term involve the use of cultural resources and modes of cultural expression grounded in everyday life experience used to articulate community voice and vision in the context of political struggle (Fischer & Brown, 1996; Ginsburg, 1997; Hofrichter, 1993). A few authors have noted the significance of cultural activism to the movement for environmental justice (Hofrichter, 1993) and others have explored the use of cultural resources in the movement by examining art, literature, theatre, and documentary film without using the term (Miller, 2000; O'Neal, 2000; Platt, 1997).
3. The author would like to thank Ms.Cromer-Campbell for her collaboration on this chapter.
4. See Parks (1965/1986).

5. Phyllis Glazer comes from a wealthy, Jewish family in Dallas, and is now married to a husband who owns one of the largest liquor distributors in the country. Media accounts of the situation in Winona often highlight the role of Glazer, a wealthy socialite prone to wearing colorful cowgirl garb, and how she has used her financial resources and flamboyant personality to pressure industry and government to do something about the health of Winona's children.

6. See http://www.mosesnonprofit.com

BIBLIOGRAPHY

36 Code of Federal Regulations. Section 219.6.

Advisory Commission on Intergovernmental Relations. (1979). *Citizen participation in the American federal system*. Washington, DC: U.S. Government Printing Office.

Alexander, N. C. (1999). *Finance for development: A dialogue with the Bretton Woods institutions*. New York: Friedrich Ebert Stiftung.

Alliance for Responsible Trade. (n.d.). *Sector analysis of the Free Trade Area of the Americas*. [On-line]. No longer available: http://www.wtowatch.org/library/admin_of_the_Free_Trade_Area_of_the_.html

Alston, D. (1990). *We speak for ourselves: Social justice, race, and environment*. Washington, DC: Panos Institute.

American Forest Congress. (1996). *Participant materials*. Courtesy of Dr. Tony Cheng, Colorado State University.

Amy, D. (1987). *The politics of environmental mediation*. New York: Columbia University Press.

Applegate, J. S. (1998). Beyond the usual suspects: The use of citizens advisory boards in environmental decision-making. *Indiana Law Journal, 73*, 903–957.

Applegate, J. S., & Sarno, D. J. (1996, Winter). Citizens get involved in cleaning up Fernald. *Forum for Applied Research and Public Policy, 11*(4), 122–124.

Applegate, J. S., & Sarno, D. J. (1997). FUTURESITE: An environmental remediation game-simulation. *Simulation and Gaming, 28*(1), 13–27.

Arnstein, S. (1969). A ladder of citizen participation. *Journal of the American Institute of Planners, 35*(4), 216–224.

Associated Press. (2001, January 9). *Coalition returns to court to ax roadless policy*. [On-line]. Available: http://www.idahostatesman.com/news/daily/20010109/LocalNews

Aune, J. A. (2001). *Selling the free market: The rhetoric of economic correctness.* New York: Guilford.

Bacow, L., & Wheeler, M. (1984). *Environmental dispute resolution.* New York: Plenum.

Bahar, R., & McCollough, G. (1996). *Laid to waste: A Chester neighborhood fights for its future.* Philadelphia, PA: DUTV-Cable 54. Videotape.

Balogh, B. (1991). *Chain reaction: Expert debate and public participation in American commercial nuclear power, 1945–1975.* Cambridge, England: Cambridge University Press.

Barber, B. (1984). *Strong democracy: Participatory politics for a new age.* Berkeley, CA: University of California Press.

Barlow, M. (2001a). *The Free Trade Area of the Americas: The threat to social programs, environmental sustainability and social justice (A special report).* San Francisco: The International Forum on Globalization.

Barlow, M. (2001b). *The future of nation-states.* Address to the World Social Forum, Porto Alegre, Brazil. [On-line]. Available: http://www.canadians.org/publications/speechnationstates-brazil.pdf

Bauman, Z. (1986–1987, Winter). The left as the counterculture of modernity. *Telos*, pp. 81–93.

Bauman, Z. (1996). From pilgrim to tourist—or a short history of identity. In S. Hall & P. du Gay (Eds.), *Questions of cultural identity* (pp. 18–36). London: Sage.

Baxter, L. A., & Montgomery, B. M. (1996). *Relating: Dialogues and dialectics.* New York: Guilford Press.

Beck, U. (1992). *Risk society: Towards a new modernity.* Thousand Oaks, CA: Sage.

Beck, U. (1994). *Ecological enlightenment: Essays on the politics of the risk society.* Atlantic Highlands, NJ: Humanities Press International.

Beck, U. (1995). *Ecological politics in an age of risk.* London: Polity.

Beck, U. (1999). *World risk society.* Cambridge, England: Polity.

Bell, M. (1977). *Savannah.* Savannah, GA: Historic Savannah Foundation.

Bello, W. (2001, March 27). Developing states resist calls for new trade talks. *Bangkok Post.* [On-line]. Available: http://scoop.bangkokpost.co.th/bkkpost/2001/march2001/bp20010327/270301_News21.html

Belsten, L. A. (1996). Environmental risk communication and community collaboration. In S. A. Muir & T. L. Veenendall (Eds.), *Earthtalk: Communication empowerment for environmental action* (pp. 27–41). Westport, CT: Praeger Press.

Benton, L. M., & Short, J. R. (1999). *Environmental discourse and practice.* Oxford, England: Blackwell.

Bernton, J., & Hogan, M. (1999, October 14). Clinton acts to protect roadless forest areas. *The Oregonian,* p. A1.

Bharucha, R. (1998). *In the name of the secular: Contemporary cultural activism in India.* New York: Oxford University Press.

Blahna, D. J., & Yonts-Shepard, S. (1989). Public involvement in resource planning: Toward bridging the gap between policy and implementation. *Society and Natural Resources, 2*(3), 209–227.

Blair, C. (1999). Challenges and openings in rethinking rhetoric: Contemporary U.S. memorial sites as exemplars of rhetoric's materiality. In J. Selzer & S. Crowley (Eds.), *Rhetorical bodies: Toward a material rhetoric* (pp. 16–57). Madison: University of Wisconsin Press.

Bohman, J. (1997). Deliberative democracy and effective social freedom. In J. Bohman & W. Rehg (Eds.), *Deliberative Democracy* (pp. 321–348). Cambridge, MA: MIT.

Bonfield, T. (2001a, August 23). Fernald study group ended over some members' protests. *Cincinnati Enquirer.* [On-line]. Available: http://enquirer.com/editions/2001/08/23/loc_fernald_study_group.html

Bonfield, T. (2001b, August 26). Questions linger over Fernald. *Cincinnati Enquirer.* [On-line]. Available: http://enquirer.com/editions/2001/08/26/loc_questions_linger.html

Bormann, E. G. (1990). *Small group communication: Theory and practice.* 3rd Ed. New York: Harper & Row.

Bourdieu, P. (1986). The forms of capital. In J. Richardson (Ed.), *Handbook of theory and research for the sociology of education* (pp. 241–258). New York: Greenwood Press. First published in German in 1983.

Brecher, J., Costello, T., & Smith, B. (2000). *Globalization from below: The power of solidarity.* Cambridge, MA: South End.

Brick, P., & Weber, E. P. (2001). Will rain follow the plow? Unearthing a new environmental movement. In P. Brick, D. Snow, & S. van de Wetering (Eds.), *Across the great divide: Explorations in collaborative conservation and the American West* (pp. 15–24). Washington, DC: Island Press.

Brown, L. D., & Fox, J. (2000). Transnational civil society coalitions and the World Bank: Lessons from project and policy influence campaigns. *IDR Reports Vol. 1, No 1.* Boston: Institute for Development Research.

Brown, P., & Mikkelsen, E. J. (1990/1997). *No safe place: Toxic waste, leukemia, and community action.* Berkeley: University of California Press.

Bruner, M., & Oelschlaeger, M. (1994). Rhetoric, environmentalism, and environmental ethics. *Environmental Ethics, 16*(4), 337–396.

Bryant, B. (1995). *Environmental justice: Issues, policies, and solutions.* Washington, DC: Island Press.

Bullard, R. D. (1990). *Dumping in Dixie: Race, class, and environmental quality.* Boulder, CO: Westview.

Bullard, R. D. (1991). *Documentary highlights of the First National People of Color Environmental Leadership Summit.* Washington, DC: United Church of Christ Commission for Racial Justice. Videotape.

Bullard, R. D. (1993). *Confronting environmental racism: Voices from the grassroots.* Boston, MA: South End Press.

Bullard, R. D. (Ed.). (1994). *Unequal protection: Environmental justice and communities of color.* San Francisco: Sierra Club Books.

Bullard, R. D., & Wright, B. H. (1987, Winter). Environmentalism and the politics of equity: Emergent trends in the black community. *Mid-American Review of Sociology, 12,* 21–37.

Burbach, R. (2001). *Globalization and postmodern politics.* Sterling, VA: Pluto.

Burgess, G., & Burgess, H. (1995). Beyond the limits: Dispute resolution of intractable environmental conflicts. In J. W. Blackburn & W. M. Bruce (Eds.), *Mediating environmental conflicts: Theory and practice* (pp. 101–119). Westport, CT: Quorum.

Burke, K. (1966). *Language as symbolic action: Essay on life, literature, and method.* Berkeley: University of California Press.

Burns, S. (2001). A civic conversation about public lands: Developing community governance. *Journal of Sustainable Forestry, 13*(1–2), 271–290.

Calhoun, C. (1992). Introduction. In C. Calhoun (Ed.), *Habermas and the public sphere* (pp. 1–50). Cambridge, MA: MIT.

Cancian, F. M., & Armstead, C. (1992). Participatory research. In E. F. Borgatta & M. L. Borgatta (Eds.), *Encyclopedia of sociology* (pp. 1427–1432). New York: Macmillan.

Cantrill, J. G. (1993). Communication and our environment: Categorizing research in environmental advocacy. *Journal of Applied Communication Research, 21*(1), 66–95.

Cantrill, J. G. (1998, November). Arguing against the "not in my neighborhood or basin!" syndrome: Using human dimensions research to preempt protected areas conflicts in the Lake Superior Basin and on the Canadian Shield. Paper presented at the National Communication Association Convention, New York, NY.

Cantrill, J. G., & Oravec, C. L. (Eds.). (1996). *The symbolic earth: Discourse and our creation of the environment.* Lexington,: University of Kentucky Press.

Carpenter, S. L. (1996). Keynote speech. Seminar on Effective Public Participation and Community Planning. Albany, NY, New York Planning Federation. n.p.

Carpenter, S. L., & Kennedy, W. J. D. (1986/1988). *Managing public disputes: A practical guide to handling conflict and reaching agreements.* San Francisco: Jossey-Bass.

CEQ. (1997). *The National Environmental Policy Act: A study of its effectiveness after twenty-five years*. Washington, DC: Government Printing Office.

Checkoway, B. (1981). The politics of public hearings. *Journal of Applied Behavioral Science, 17*(4), 566–581.

Chess, C. (1995). *Public participation and the environment: What works?* Unpublished paper, State University of New York College of Environmental Science & Forestry, Syracuse, New York.

Chess, C., & Purcell, K. (1999). Public participation and the environment: Do we know what works? *Environmental Science & Technology, 33*(16), 2685–2692.

Chiriboga, M. (2000). *NGOs and the World Bank: Lessons and challenges.* Montreal, Canada: Montreal International Forum. [On-line]. Available: http://www.fimcivilsociety.org/etud_chiriboga_english_text.html

Church, D. (1995, March 20). (Director of the New York State Planning Federation). *Alternative dispute resolution in land use process.* Presentation at Albany Law School, Albany, New York.

CIMSLP. (1979). Public involvement in maritime facility development. Washington, DC: National Academy of Sciences.

Cispus Adaptive Management Area. (1994, Summer). *The Cispus AMA Quarterly.* Randle, WA: Gifford Pinchot National Forest.

Clark, G., Halloran, S. M., & Woodford, A. (1996). Thomas Cole's vision of "nature" and the conquest theme in American culture. In C. G. Herndl, & S. C. Brown (Eds.), *Green culture: Environmental rhetoric in contemporary America* (pp. 261–280). Madison: University of Wisconsin Press.

Clark, R. N., Stankey, G. H., & Shannon, M. A. (1999). The social component of the Forest Ecosystem Management Assessment Team (FEMAT). In H. K. Cordell & J. C. Bergstrom (Eds.), *Integrating social sciences with ecosystem management: Human dimensions in assessment, policy, and management* (pp. 237–264). Champaign, IL: Sagamore Publishing.

Clary, B. B., & Hornney, R. (1995). Evaluating ADR as an approach to citizen participation in siting a low-level nuclear waste facility. In J. W. Blackburn, & W. M. Bruce (Eds.), *Mediating environmental conflicts: Theory and practice* (pp. 121–138). Westport, CT: Quorum.

Cleland, L. (1996). *Seventh American Forest Congress. Alliance For America.* [On-line]. Available: http://www.allianceforamerica.org/Oldweb/forcong.html

Clinton, W. J. (1999a). *Memorandum for the Secretary of Agriculture.* USDA Forest Service Roadless Conservation Web site. [On-line]. Available: http://www.roadless.fs.fed.us/documents

Clinton, W. J. (1999b). *Remarks by the President at "Roadless" lands event.* USDA Forest Service Roadless Conservation Web site. [On-line]. Available: http://www.roadless.fs.fed.us/documents

Clinton, W., & Gore, A. (1995, March 16). *Reinventing environmental regulation*. [On-line]. Available: http://www.npr.gov/library/rsreport/25la.html

Cochran, C. (1998, December 30). Water cap may dry up county's residential growth. *Savannah Morning Electronic Edition News*. [On-line]. Available: http://www.savannahnow.com

Cohen, J. (1997). Deliberation and democratic legitimacy. In J. Bohman & W. Rehg (Eds.), *Deliberative democracy* (pp. 67–92). Cambridge, MA: MIT.

Cohen, J. L., & Arato, A. (1994). *Civil society and political theory.* Cambridge, MA: MIT Press.

Cole, R. L., & Caputo, D. A. (1984, June). The public hearing as an effective citizen participation mechanism: A case study of the General Revenue Sharing Program. *The American Political Science Review, 78,* 404–416.

Coleman, J. (1990). *Foundations of social theory.* Cambridge, MA: Harvard University Press.

Collingsridge, D., & Reeve, C. (1986). *Science speaks to power: The role of experts in policymaking*. New York: St. Martins.

Collins-Jarvis, L. (1997). Participation and consensus in collective action organizations: The influence of interpersonal versus mass-mediated channels. *Journal of Applied Communication Research, 25,* 1–16.

Commission on Global Governance. (1995). *Our global neighborhood: The report of the Commission on Global Governance.* Oxford, England: Oxford University Press.

Compa, L. (1998). Democratizing the trade debate. *Foreign Policy in Focus 3*(23), 1–7. [On-line]. Available: http://www.foreignpolicy-infocus.org/briefs/vol3/v3n23trad_body.html

Conca, K. (1996). Greening the UN: Environmental organizations and the UN system. In T. G. Weiss & L. Gordenker, Jr. (Eds.), *NGOs, the UN, and global governance* (pp.103–120). Boulder, CO: Lynne Rienner.

Condit, C. M. (1987a). Crafting virtue: The rhetorical construction of public morality. *Quarterly Journal of Speech, 73*(1), 79–97.

Condit, C. M. (1987b). Democracy and civil rights: The universalizing influence of public argumentation. *Communication Monographs, 54,* 1–18.

Condit, C. M. & Lucaites, J. L. (1993). *Crafting equality: America's Anglo-African word.* Chicago: University of Chicago Press.

Conflict history. (1998, June 14). *The Savannah Morning News Electronic Edition*. [On-line]. Available: http://www.savannahnow.com

Cooper, M. M. (1996). Environmental rhetoric in the age of hegemonic politics: Earth First! and the Nature Conservancy. In C. G. Herndl & S. C. Brown (Eds.), *Green culture: Environmental rhetoric in contemporary America* (pp. 236–260). Madison: University of Wisconsin Press.

Cortner, H. J., & Schweitzer, D. L. (1981). Institutional limits to national public planning for forest resources: The Resource Planning Act. *Natural Resources Journal, 2*(2), 203–322.

COSMOS. (2001a). *Final report for the evaluation of the health effects subcommittee advisory process: Executive summary*. Bethesda, MD: COSMOS Corporation.

COSMOS. (2001b). *Final report for the evaluation of the health effects subcommittee advisory process: Volume I*. Bethesda, MD: COSMOS Corporation.

COSMOS. (2001c). *Final report for the evaluation of the health effects subcommittee advisory process: Volume II*. Bethesda, MD: COSMOS Corporation.

Cox, R. W. (1992). Global *perestroika*. In R. Miliband & L. Panitch (Eds.), *Socialist register* (pp. 26–43). London: Merlin.

Crable, R. E. (1990). "Organizational rhetoric" as the fourth great system: Theoretical, critical, and pragmatic implications. *Journal of Applied Communication Research, 18,* 115–128.

Cronen, V. E. (1995). Practical theory and the tasks ahead for social approaches to communication. In L. Leeds-Hurwitz (Ed.), *Social approaches to communication* (pp. 217–242). New York: Guilford Press.

Cronen, V. E. (2000, January 27). *Practical theory and a naturalistic account of inquiry.* Keynote presentation, Baylor University Conference on Practical Theory, Public Participation, and Community, Waco, TX.

Cronen, V. E. (2001). Practical theory, practical art, and the pragmatic-systemic account of inquiry. *Communication Theory, 11,* 14–35.

Crosby, N. (1995). Citizens juries: One solution for difficult environmental problems. In O. Renn, T. Webler, & P. Wiedermann (Eds.), *Fairness and competence in citizen participation: Evaluating models for environmental discourse* (pp. 157–174). Boston, MA: Kluwer Academic Press.

Crotty, W. (Ed.). (1991). *Political participation and American democracy.* New York: Greenwood Press.

Crowfoot, J. E., & Wondolleck, J. M. (1990). *Environmental disputes: Community involvement in conflict resolution.* Washington, DC: Island Press.

Curwood, S. (2001, April 9). Interview with Mike Dombeck, former USFS chief. *Living on Earth.* National Public Radio. [On-line]. Available: http://www.loe.org

Dahlberg, K. A., Soroos, M. S., Feraru, A. T., Harf, J. E., & Trout, B. T. (1985). *Environment and the global arena.* Durham, NC: Duke University.

Daniels, C. L., & Schutt, J. A. (1999). *Planning for the protection and management of the cultural resources on the Placitas open space.* (Research report). Placitas, NM: Las Placitas Association.

Daniels, S. E., & Walker, G. B. (1994, April 21). *Searching for effective natural resource policy: The special challenges of ecosystem management.* Paper presented at the Natural Resources Week Symposium on Ecosystem Management, Utah State University, Logan, Utah.

Daniels, S. E., & Walker, G. B. (1995, February 23–24). *Collaborative learning workshop* (materials packer). Prepared for Randle and Packwood Ranger Districts, Gifford Pinchot National Forest, USDA Forest Service.

Daniels, S. E., & Walker, G. B. (1996). Collaborative learning: Improving public deliberation in ecosystem-based management. *Environmental Impact Statement Assessment Review, 16,* 71–102.

Daniels, S. E., & Walker, G. B. (1999). Rethinking public participation in natural resource management: Concepts from pluralism and five emerging approaches. In J. Anderson (Ed.), *Pluralism and sustainable forestry and rural development: Proceedings of an international workshop* (pp. 29–48). Rome, Italy: Food and Agriculture Organization [FAO] of the United Nations.

Daniels, S. E., & Walker, G. B. (2001). *Working through environmental conflict: The collaborative learning approach.* Westport, CT: Praeger.

Daniels, S. E., Walker, G. B., Carroll, M. S., & Blatner, K. A. (1996). Using collaborative learning in fire recovery planning. *Journal of Forestry, 94*(8), 4–9.

Davis, S. G. (1997). *Spectacular nature: Corporate culture and the Sea World experience.* Berkeley: University of California Press.

Delicath, J. W. (2001). Re-invigorating the public sphere, reviving democratic praxis: Environmental justice and the demand for public participation. In C. B. Short & D. H. Short (Eds.), *Proceedings of the Fifth Biennial Conference on Communication and Environment* (pp. 32–44). Flagstaff: Northern Arizona University.

DeLuca, K. M. (1999). *Image politics: The new rhetoric of environmental activism.* New York: Guilford.

DeLuca, K. M., & Demo, A. T. (2000, September). Imaging nature: Watkins, Yosemite, and the birth of environmentalism. *Critical Studies in Media Communication, 17*(3), 241–260.

DePalma, A. (2001, March 11). NAFTA's powerful little secret. *The New York Times* [Money and Business Section], pp. 1, 13.

DeSario, J., & Langton, S. (1987a). Toward a metapolicy for social planning. In J. DeSario & S. Langton (Eds.), *Citizen participation in public decision making* (pp. 205–222). Westport, CT: Greenwood Press.

DeSario, J., & Langton, S. (1987b). Citizen participation and technocracy. In J. DeSario & S. Langton (Eds.), *Citizen participation in public decision making* (pp. 3–18). Westport, CT: Greenwood Press.

DeSario, J., & Langton, S. (Eds.). (1987c). *Citizen participation in public decision making.* New York: Greenwood Press.

Devine, O. J., Qualters, J. R., Morrissey, J. L., & Wall, P. A. (1998). *Estimation of the impact of the former Feed Materials Production Center (FMPC) on lung cancer mortality in the surrounding community.* Atlanta, GA: Centers for Disease Control.

Devine, O. J., Qualters, J. R., Morrissey, J. L., & Wall, P. A. (2000). *Screening level estimates of the lifetime risk of developing: Kidney cancer, female breast cancer, bone cancer, leukemia.* Atlanta, GA: Centers for Disease Control.

DeWitt, J. (1994). *Civic environmentalism: Alternatives to regulation in states and communities.* Washington, DC: Congressional Quarterly Press.

Di Chiro, G. (1996). Nature as community: The convergence of environment and social justice. In W. Cronon (Ed.), *Uncommon ground: Rethinking the human place in nature* (pp. 298–320). New York: W.W. Norton & Co.

Di Chiro, G. (1998). Environmental justice from the grassroots: Reflections on history, gender, and expertise. In D. Faber (Ed.), *The struggle for ecological democracy: Environmental justice movements in the United States* (pp. 104–136). New York: Guilford Press.

Di Chiro, G. (2000). Bearing witness or taking action? Toxic tourism and environmental justice. In R. Hofrichter (Ed.), *Reclaiming the environmental debate: The politics of health in a toxic culture* (pp. 275–300). Cambridge, MA: MIT Press.

Douglas, M. (1968). *Purity and danger: An analysis of concepts of pollution and taboo.* New York: Frederick A. Praeger.

Dryzek, J. S. (1990). *Discursive democracy.* Cambridge: Cambridge University Press.

Dryzek, J. S. (1996). Strategies of ecological democratization. In W. M. Lafferty & J. Meadowcroft (Eds.), *Democracy and the environment: Problems and prospects* (pp. 108–123). Cheltenham, England: Edward Elgar.

Duffield, J. J., & Depoe, S. P. (1997, February 21). Lessons from Fernald: Reversing NIMBYism through democratic decision-making. *EPA Risk Policy Report, 3,* 31–34.

Dukes, E. F. (1996). *Resolving public conflict: Transforming community and governance.* New York: Manchester University Press.

Dunham, H. W. (1986). The community today: Place or process? *Journal of Community Psychology, 14*(4), 399–404.

Dunkley, G. (2000). *The free trade adventure: The WTO, the Uruguay round and globalism.* London: Zed Books.

Dunne, N. (2001, July 4). Critics attack timing of Americas free trade draft. *Financial Times* (USA ed. 2), [International Economy Section], p. 4.

Eccleston, C. H. (1999). *The NEPA planning process: A comprehensive guide with emphasis on efficiency.* New York: John Wiley & Sons.

Edelman, M. (1995). *From art to politics: How artistic creations shape political conceptions.* Chicago: University of Chicago Press.

Edgerton, R. (1995, September). "Bowling alone": An interview with Robert Putnam about American's collapsing civic life. *AAHE Bulletin* [American Association of Higher Education], 1; 3. [On-line]. Available: http://muse.jhu.edu/demo/journal_of_democracy/v006/putnam.interview.htm

Editorial: Politics and water. (1998, June 14). *Savannah Morning News Electronic Edition* [On-line]. Available: http://www.savannahnow.com

Editorial: Project should proceed. (1998, July 21). *The Savannah Morning News Electronic Edition.* [On-line]. Available: http://www.savannahnow.com

Environmental Defense (2000, December 15). *About scorecard/What's new.* [On-line]. Available: http://www.scorecard.org/about/txt/new.html#volunteer

Fairfax, S. K. (1975). Public involvement and the Forest Service. *Journal of Forestry, 73*(10), 657–659.

Falk, R. (1999). *Predatory globalization: A critique.* Cambridge, England: Polity.

Fancher, B. (1976). *Savannah: A renaissance of the heart.* Garden City, NY: Doubleday.

Farrell, R. T. (1999). Aiming for excellence: EPA's commitment to innovation. *ECOStates.* [On-line]. Available: http://www.epa.gov

Farrell, T. (1993). *Norms of rhetorical culture.* New Haven, CT: Yale University Press.

Farrell, T. B., & Goodnight, G. T. (1981). Accidental rhetoric: The root metaphors of Three-Mile Island. *Communication Monographs, 48,* 271–300.

FCAB. (2002). *Advisory Board recommendations.* [On-line]. Available: http://www.fernaldcab.org/recomandreports.html

FCTF. (1995, July). *Fernald Citizens Task Force: Recommendations on remediation levels, waste disposition, priorities, and future use.* Fernald, Ohio.

Feinberg, R. E., & Rosenberg, R. L. (Eds.). (1999). *Civil society and the Summit of the Americas: The 1998 Santiago summit.* Miami, FL: University of Miami Press.

Felski, R. (1994). Does the concept of feminist aesthetics facilitate or inhibit cultural activism? In P. Goldstein (Ed.), *Styles of cultural activism: From theory and pedagogy to women, Indians, and communism* (pp. 203–215). Newark: University of Delaware Press.

FEMAT. (1993). *Forest ecosystem management: An ecological, economic, and social assessment.* Washington, DC: U.S. Government Printing Office, 1993-793-071.

FERMCO. (1995, July). *Fernald Environmental Management Project 1994 site environmental report.* Fernald, Ohio.

Fetterman, D. M. (1998). *Ethnography: Step by step.* Thousand Oaks, CA: Sage.

FHES. (1996, November). *Draft mission statement of Fernald Health Effects Subcommittee.* Fernald, OH: CDC.

Field, N. (1995). Cultural activism. In N. Field (Ed.), *Over the rainbow: Money, class, and homophobia* (pp. 120–132). London: Pluto Press.

Field, R. C. (1998). Risk and justice: Capitalist production and the environment. In D. Faber (Ed.), *The struggle for ecological democracy: Environmental justice movements in the United States* (pp. 81–103). New York: Guilford.

Fine, B. (2000). Bringing the social back into economics: Progress or reductionism? *Research Paper No. 731.* The University of Melbourne Department of Economics.

Fine, B. (2001). *Social capital versus social theory: Political economy and social science at the turn of the millennium.* London and New York: Routledge.

Fiorino, D. J. (1989a). Environmental risk and democratic process: A critical review. *Columbia Journal of Environmental Law, 14*(2), 501–547.

Fiorino, D. J. (1989b). Technical and democratic values in risk analysis. *Risk Analysis, 9*(3), 293–299.

Fiorino, D. J. (1990, Spring). Citizen participation and environmental risk: A survey of institutional mechanisms. *Science, Technology, & Human Values, 15*(2), 226–243.

Fiorino, D. J. (1996). Environmental policy and the participation gap. In W. M. Lafferty & J. Meadowcroft (Eds.), *Democracy and the environment: Problems and prospects* (pp. 194–212). Cheltenham, England: Edward Elgar.

Fischer, E. F., & Brown, R. M. (Eds.). (1996). *Maya cultural activism in Guatemala.* Austin: University of Texas Press.

Fischer, F. (2000). *Citizens, experts, and environment: The politics of local knowledge.* Durham, NC: Duke University Press.

Fischer, F., & Forester, J. (Eds.). (1993). *The argumentative turn in policy analysis and planning.* Durham, NC: Duke University Press.

Fisher, W. R. (1987). *Human communication as narration: Toward a philosophy of reason, value, and action.* Columbia: University of South Carolina Press.

Fisher, W. R. (1994). A case: Public moral argument. In R. Anderson, K. N. Cissna, & R. C. Arnett (Eds.), *The reach of dialogue: Confirmation, voice, and community* (pp. 173–177). Cresskill, NJ: Hampton.

Forest Congress Information Center (1998). *Seventh American Forest Congress: Executive summary.* Yale Forest Forum. [On-line]. Available: http://www.yale.edu/forest_congress/summary/sumexecutive.html

Fraser, N. (1992). Rethinking the public sphere: A contribution to the critique of actual existing democracy. In C. Calhoun (Ed.), *Habermas and the public sphere* (pp. 109–142). Cambridge, MA: MIT.

Friedman, T. L. (1999). *The Lexus and the olive tree*. New York: Farrar, Straus and Giroux.

Friedman, T. L. (2000). *The Lexus and the olive tree: Understanding globalization* (Rev. ed.). New York: Anchor Books.

FTAA. (2001). *Free Trade Area of the Americas*. [On-line]. Available: http://www.ftaa-alca-org/SPCOMM/COMMCS_E.asp

Fukuyama, F. (1989, Summer). The end of history. *The National Interest, 16,* 3–18.

Furnish, J. R. (2000, July 26). *Statement of James R. Furnish,* Deputy Chief, National Forest System, Forest Service, United States Department of Agriculture, before the Subcommittee on Forests and Public Lands Management, Committee on Energy and Natural Resources, United States Senate, regarding the Forest Service Roadless Area Conservation Proposal. [On-line]. Available: http://www.roadless.fs.fed.us/documents

Gariepy, M. (1991). Toward a dual-influence system: Assessing the effects of public participation in environmental impact assessment for Hydro-Quebec projects. *Environmental Impact Assessment Review, 11,* 352–374.

Gates, C. T. (1998, October 3). *Keynote Address.* Annual Conference and Workshops of the International Association for Public Participation, Tempe, AZ. Videotape.

Gaventa, J. (1999). Citizen knowledge, citizen competence, and democracy building. In S. L. Elkin & K. E. Soltan (Eds.), *Citizen competence and democratic institutions* (pp. 49–65). University Park, PA: Penn State University Press.

Geertz, C. (1973). *The interpretation of cultures.* New York: Basic Books.

Gençkaya, Ö. F. (2000). States and non-state actors in environmental policy making: An overview of the GEF-BSEP NGO forum. In G. D. Dabelko & S. van Deever (Eds.), *Protecting seas: Developing capacity and fostering environmental cooperation in Europe* (pp. 74–102). Washington, DC: WWC Publication.

Gericke, K. L., & Sullivan, J. (1994). Public participation and appeals of Forest Service plans: An empirical examination. *Society and Natural Resources, 7*(2), 125–135.

Gericke, K. L., Sullivan, J., & Wellman, J. D. (1992). Public participation in national forest planning: Perspectives, procedures, and costs. *Journal of Forestry, 90*(2), 35–38.

Giddens, A. (1984). *The constitution of society: Outline of the theory of structuration.* Berkeley: University of California Press.

Giddens, A. (1990). *The consequences of modernity.* Stanford, CA: Stanford University Press.

Gieryn, T. F. (1995). Boundaries of science. In S. Jasanoff, G. E. Markle, J. C. Petersen, & T. Pinch (Eds.), *Handbook of science and technology studies* (pp. 393–443). Thousand Oaks, CA: Sage.

Ginsburg, F. (1997). "From little things, big things grow": Indigenous media and cultural activism. In R. G. Fox & O. Starn (Eds.), *Between resistance and revolution: Cultural politics and social protest* (pp. 118–144). New Brunswick, NJ: Rutgers University Press.

Gladden, J. N. (1990). *The Boundary Waters Canoe Area: Wilderness values and motorized recreation.* Ames: Iowa State University Press.

Glaser, B. L., & Strauss, A. L. (1967). *The discovery of grounded theory: Strategies for qualitative research.* Chicago: Aldine.

Glover, K. (2000). Environmental discourse and Bakhtinian dialogue: Toward a dialogic rhetoric of diversity. In N. W. Coppola, & B. Karis (Eds.), *Technical communication, deliberative rhetoric, and environmental discourse: Connections and directions* (pp. 37–54). Stamford, CT: Ablex.

Goetz, K. (2000, November 4). Vision for Fernald: Learning center. *Cincinnati Enquirer.* [On-line]. Available: http://enquirer.com/editions/2000/11/04/loc_vision_for_fernald.html

Goldenberg, S., & Frideres, J. S. (1986). Measuring the effects of public participation programs. *Environmental Impact Assessment Review, 6*(3), 273–281.

Goldman, R., & Papson, S. (1996). *Sign wars: The cluttered landscape of advertising.* New York: Guilford Press.

Goldstein, P. (Ed.). (1994). *Styles of cultural activism: From theory and pedagogy to women, Indians, and communism.* Newark: University of Delaware Press.

Goodnight, G. T. (1982). The personal, technical, and public spheres of argument: A speculative inquiry into the art of public deliberation. *Journal of the American Forensic Association, 18,* 214–227.

Goodnight, G. T., & Hingstman, D. B. (1997). Studies in the public sphere. *Quarterly Journal of Speech, 83,* 351–399.

Gordenker, L., & Weiss, T. G. (1995, September). NGO participation in the international policy process. *Third World Quarterly, 16(*3), 543–555.

Gottlieb, R. (1993). *Forcing the spring: The transformation of the American environmental movement.* Washington, DC: Island Press.

GPA. (1998a, July). *Savannah Harbor expansion feasibility study report.* Savannah, GA: GPA Public Affairs Division.

GPA. (1998b, December). *Public meeting slides.* [On-line]. Available: http://www.sysconn.com/harbor

GPA. (2000). The harbor deepening planning process. *Harbor Expansion Questions & Answers* [On-line]. Available: http://www.gaports.com

Graham, A. C. (1996). *Planning as public dialog: A social communication perspective on participatory processes.* Unpublished master's thesis. University of Washington, Seattle, WA.

Graham, A. C. (1997). A social communication analysis of public participation: The case of the Cispus Adaptive Management Area. In S. L. Senecah (Ed.), *Proceedings of the Fourth Biennial Conference on Communication and Environment* (pp. 226–243). Syracuse: State University of New York College of Environmental Science and Forestry.

Granger, M. L. (1968). *Savannah harbor: Its origin and development, 1733–1890.* Savannah, Georgia: U. S. Army Engineers District, Savannah, Corps of Engineers.

Granovetter, M. (1985). Economic action and social structure: The problem of embeddedness. *American Journal of Sociology 91*(3), 481–510.

Gray, G., & Kusel, J. (1998). Changing the rules. *American Forests, 103*(4), 27–31.

Greenpeace. (1999a). *Notes from the road: Monday, June 28.* [On-line]. Available: http://www.greenpeaceusa.org/toxics/bustour28text.htm

Greenpeace. (1999b, June 22). *With public's 'Right-to-know' in jeopardy, Greenpeace kicks off bus tour of Louisiana's worst chemical 'hot spots.'* Press release. [On-line]. Available: http://www.greenpeaceusa.org/media/press_releases/99_6_22text.htm

Grossman, K. (1994). The people of color environmental summit. In R. D. Bullard (Ed.), *Unequal protection: Environmental justice and communities of color* (pp. 272–297). San Francisco: Sierra Club Books.

Group blasts Voinovich for state's toxic waste. (1992, May 16). *The Columbus Dispatch,* p. 8B.

Gunderson, A. G. (1995). *The environmental promise of democratic deliberation.* Madison: University of Wisconsin Press.

Gutmann, A., & Thompson, D. (1996). *Democracy and disagreement.* Cambridge, MA: Belknap Press of Harvard University.

Habermas, J. (1975). *Legitimation crisis.* Boston: Beacon.

Habermas, J. (1984). *Theory of communicative action: Reason and the rationalization of society,* Vol. 1 (T. McCarthy, Trans.). Boston: Beacon Press.

Habermas, J. (1991/1962). *The structural transformation of the public sphere: An inquiry into a category of burgeois society* (T. Berger, Trans.). Cambridge, MA: MIT Press.

Hamilton, J. D. (1997). Variability of stakeholder views about citizen participation in the Fernald radium debate. In S. Senecah (Ed.), *Proceedings of the 4th Biennial Conference on Communication and our Environment* (pp. 254–264). Syracuse: State University of New York College of Environmental Science and Forestry.

Hance, G. J., Chess, C., & Sandman, P. M. (1995). *Improving dialogue with communities: A risk communication manual for government.* Trenton: New Jersey Department of Environmental Protection.

Hardin, R. (1999). Deliberation: Method, not theory. In S. Macedo (Ed.), *Deliberative politics: Essays on democracy and disagreement* (pp. 103–119). New York: Oxford University Press.

Hardy-Short, D. C., & Short, C. B. (1995, Spring). Fire, death, and rebirth: A metaphoric analysis of the 1988 Yellowstone fire debate. *Western Journal of Communication, 59*, 103–125.

Harless, J. D. (1992). Local government environmental advisory boards. *National Civic Review, 81,* 9–18.

Healey, P. (1992). Planning through debate: The communicative turn in planning theory. *TPR, 63*(2), 143–162.

Healey, P. (1993). Planning through debate: The communicative turn in planning theory. In F. Fischer & J. Forester (Eds.), *The argumentative turn in policy analysis and planning* (pp. 233–253). Durham, NC: Duke University Press.

Heberlein, T. A. (1976, January). Some observations on alternative mechanisms for public involvement: The hearing, public opinion poll, the workshop, and the quasi-experiment. *Natural Resources, 16*(1), 197–212.

Held, D., & McGrew, A. (2000). The great globalization debate. In D. Held & A. McGrew (Eds.), *The global transformations reader: An introduction to the globalization debate* (pp. 1–45). Cambridge, England: Polity.

Hemispheric Social Alliance. (2000, November 7). Letter to Dr. Adalberto Rodriguez Giavarini, Chair, FTAA Negotiating Committee.

Hendee, J., Lucas, R., Tracy, R., Staed, T., Clark, R., & Stankey, G. (1976). In J. Pierce, & H. Doerksen (Eds.), *Water politics and public involvement* (pp. 125–144.) Ann Arbor, MI: Ann Arbor Science.

Henning, D. H. (1974). *Environmental policy and administration.* New York: American Elsevier.

Herndl, C. G., & Brown, S. C. (1996). *Green culture: Environmental rhetoric in contemporary America*. Madison: University of Wisconsin Press.

Hewitt, B., & Calkins, L. B. (1998, January 22). Chemical warfare. *People Weekly, 49*, 52–55.

Higgins, R. R. (1994, Fall). Race, pollution, & the mastery of nature. *Environmental Ethics, 16,* 251–264.

Hofrichter, R. (1993). Cultural activism and environmental justice. In R. Hofrichter (Ed.), *Toxic struggles: The theory and practice of environmental justice* (pp. 85–96). Philadelphia: New Society Publishers.

Hofrichter, R. (2000). Introduction: Critical perspectives on human health and the environment. In R. Hofrichter (Ed.), *Reclaiming the environmental debate: The politics of health in a toxic culture* (pp. 1–15). Cambridge, MA: MIT Press.

Hogan, M. (1999, 16 October). Clinton draws criticism with latest forest plan. *The Oregonian,* p. A7.

Hohendahl, P. U. (1992). The public sphere: Models and boundaries. In C. Calhoun (Ed.), *Habermas and the public sphere* (pp. 99–108). Cambridge, MA: MIT.

Hollandsworth, S. (1997, September). Phyllis Glazer: Once she was a Dallas housewife. Now she's the Toxic Avenger. *Texas Monthly,* pp. 133, 166.

Hughes, J. (1999, November 21). Critics say roadless plan goes too far. *The Oregonian,* pp. A3, A7.

Huner, J. (1998, October 29–30). *Environmental regulation and international investment agreements: Lessons from the MAI, introduction.* Seminar Trade, Investment and the Environment, RIIA. London: Chatham House.

International Institute for Sustainable Development. (2001). *Private rights, public problems.* Winnipeg, Canada. [On-line]. Available from the International Institute for Sustainable Development http://www.iisd.ca/about/prodcat/ordering.html

John, D., & Mlay, M. (1999). Community-based environmental protection: Encouraging civic environmentalism. In K. Sexton, A. A. Marcus, K. W. Easter, & T. D. Burkhardt (Eds.), *Better environmental decisions: Strategies for governments, businesses and communities* (pp. 353–376). Washington, DC: Island Press.

Johnson, T. (1998). *The second creation story: Redefining the bond between religion & ecology.* Sierra Club Web site. [On-line]. Available: http://www.sierraclub.org/sierra/199811/second.html

Jones, E. S., & Callaway, W. (1995). Neutral bystander, intrusive micromanager, or useful catalyst? The role of Congress in effecting change within the Forest Service. *Policy Studies Journal, 23*(2), 337–350.

Jones, E. S., & Taylor, C. P. (1995). Litigating agency change: The impact of the courts and administrative appeals process on the Forest Service. *Policy Studies Journal, 23*(2), 310–336.

Jordan, J. (1982/1985, 1993). Report from the Bahamas. In W. Brown & A. Ling (Eds.), *Visions of America: Personal narratives from the promised land* (pp. 305–315). New York: Persea Books.

Joss, S., & Durant, J. (Eds.). (1995). *Public participation in science: The role of consensus conferences in Europe.* London: Science Museum.

Journal of Advertising, XXIV.2. (1995, September). Special Issue on Green Advertising.

Juanillo, N. K., & Scherer, C. W. (1995). Attaining a state of informed judgments: Toward a dialectical discourse on risk. In B. Burleson (Ed.), *Communication Yearbook, 18* (pp. 278–299). Thousand Oaks, CA: Sage Publications.

Judis, J. B. (2000, May 22). Fix it or nix it. *The American prospect 11*(13). [Online]. Available: http://www.prospect.org/print/V11/13/judis-j.html

Kaminstein, D. S. (1996). Persuasion in a toxic community: Rhetorical aspects of public meetings. *Human Organization, 55*(4), 458–464.

Kaplan, C. (1996). *Questions of travel: Postmodern discourses of displacement.* Durham, NC: Duke University Press.

Kasperson, R. (1986). Six propositions on public participation and their relevance for risk communication. *Risk Analysis, 6,* 275–281.

Katz, S. B., & Miller, C. R. (1996). The low-level radioactive waste citing controversy in North Carolina: Toward a rhetorical model of risk communication. In C. G. Herndl & S. C. Brown (Eds.), *Green culture: Environmental rhetoric in contemporary America* (pp. 111–140). Madison: The University of Wisconsin Press.

Kemmis, D. (1995). *The good city and the good life: Renewing the sense of community.* New York: Houghton Mifflin.

Kessler, W. B. (1991). "New perspectives" and the Forest Service. *Journal of Soil and Water Conservation, 46*(1), 6–7.

Keyes, C. (1995, January 31). *Environmental justice.* Columbia, MI: Jesus People Against Pollution. Videotape.

Keyes, R. (1973). *We the lonely people: Search for community.* New York: Harper & Row.

Keystone Center. (1993, February). *Interim report of the Federal Facilities Environmental Restoration Dialogue Committee: Recommendations for improving the federal facilities environmental restoration decision-making and priority-setting processes and setting priorities in the event of funding shortfalls.* Washington, DC. U.S. Government Printing Office.

Keystone Center. (1996, April). *Final report of the Federal Facilities Environmental Restoration Dialogue Committee: Consensus principles and recommendations for improving federal facilities cleanup.* Washington, DC: U.S. Government Printing Office.

Killingsworth, M. J., & Palmer, J. S. (1992). *Ecospeak: Rhetoric and Environmental Politics in America.* Carbondale: Southern Illinois University Press.

Killingsworth, M. J., & Palmer, J. S. (1995). The discourse of "environmentalist hysteria." *Quarterly Journal of Speech, 81*(1), 1–19.

Killingsworth, M. J., & Palmer, J. S. (1996). Millennial ecology: The apocalyptic narrative from Silent Spring to global warming. In C. G. Herndl & S. C. Brown (Eds.), *Green culture: Environmental rhetoric in contemporary America* (pp. 21–45). Madison: University of Wisconsin Press.

Kim, K. W. (2000). Trade, monetary policy, and democracy. In M. F. Plattner & A. Smolar (Eds.), *Globalization, power, and democracy* (pp. 117–132). Baltimore: Johns Hopkins University Press.

Kinsella, W. J. (2001). Nuclear boundaries: Material and discursive containment at the Hanford nuclear reservation. *Science as Culture, 10*(2), 163–194.

Kleinmann, D. L. (Ed.). (2000). *Science, technology, and democracy*. Albany: State University of New York Press.

Knight, J., & Johnson, J. (1997). What sort of equality does deliberative democracy require? In J. Bohman & W. Rehg (Eds.), *Deliberative democracy* (pp. 279–320). Cambridge, MA: MIT.

Kobrin, S. J. (1998, Fall). The MAI and the clash of globalizations. *Foreign Policy 112,* 97–109.

Koivisto, J., & Valiverronen, E. (1996). The resurgence of critical theories of the public sphere. *Journal of Communication Inquiry, 20,* 18–36.

Kolb, D. M. & Associates. (1994). *When talk works: Profiles of mediators*. San Francisco: Jossey-Bass.

Korten, D. C. (1999). *The post-corporate world: Life after capitalism.* San Francisco and West Hartford, CT: Berrett-Koehler Publishers and Kumarian Press.

Kraft, M. E. (1999). Making decisions about environmental policy. In K. Sexton, A. A. Marcus, K. W. Easter, & T. D. Burkhardt (Eds.), *Better environmental decisions: Strategies for governments, businesses, and communities* (pp.15–36). Washington, DC: Island Press.

Kraft, M. E., & Wuertz, D. (1996). Environmental advocacy in the corridors of government. In J. G. Cantrill, & C. L. Oravec (Eds.), *The symbolic earth: Discourse and our creation of the environment* (pp. 95–122). Lexington: University Press of Kentucky.

Kreuzwieser, M. (1995, May/June). Washing away? *Savannah Magazine, 6,* 29–35.

Krimsky, S. (1992). The role of theory in risk studies. In S. Krimsky, & D. Golding (Eds.), *Social theories of risk* (pp. 3–22). Westport, CT: Praeger.

Krueger, G. (1998a, June 14). River deep, rift wide. *Savannah Morning News Electronic Edition.* [On-line]. Available: http://www.savannahnow.com

Krueger, G. (1998b, June 16). A question of balance. *Savannah Morning News Electronic Edition.* [On-line]. Available: http://www.savannahnow.com

Kruger, L. E. (1996). *Understanding place as a cultural system: Implications of theory and method.* Unpublished doctoral dissertation. University of Washington, Seattle, WA.

Krugman, P. (1999). *The return of depression economics*. London: Allen Lane, Penguin Press.

Kuletz, V. L. (1998). *The tainted desert: Environmental and social ruin in the American west.* New York: Routledge.

Kweit, M. G., & Kweit, R. W. (1984). The politics of policy analysis: The role of citizen participation in analytic decision-making. *Policy Studies Review 3*(2), 234–245.

Laird, F. N. (1989). The decline of deference: The political context of risk communication. *Risk Analysis, 9*(4), 543–550.

Laird, F. N. (1993). Participatory analysis, democracy, and technological decision making. *Science, Technology, & Human Values, 18*(3), 341–361.

Lance, A. G. (2001). *Attorney General to move forward with Roadless lawsuit.* News release, State of Idaho Office of the Attorney General. [On-line]. Available: http://www.2state.id.us/ag/newsrel/2001

Landais-Stamp, P. (1991). The mediation of environmental conflicts. In T. Woodhouse (Ed.), *Peacemaking in a troubled world* (pp. 266–288). New York: St. Martin's Press.

Lange, J. I. (1990, Fall). Refusal to compromise: The case of Earth First! *Western Journal of Speech Communication, 54,* 473–494.

Lange, J. I. (1993). The logic of competing information campaigns: Conflict over old growth and the spotted owl. *Communication Monographs, 60,* 239–257.

Langton, S. (1978). *Citizen participation in America: Essays on the state of the art.* Lexington, MA: Lexington Books.

Las Placitas Association. (1999). *General Information* [Brochure]. Placitas, NM: Author.

Lavelle, M., & Coyle, M. (1992, September 21). Critical mass builds on environmental equity. *The National Law Journal.* Washington Briefs, Section 5.

Lawrence, R. L., & Daniels, S. E. (1996, October). Public involvement in natural resource decision making: Goals, methodology, and evaluation. *Papers in Forest Policy, 3.* Corvallis: Forest Research Laboratory, Oregon State University.

Leeds-Hurwitz, W. (1995). Introducing social approaches. In W. Leeds-Hurwitz (Ed.), *Social approaches to communication* (pp. 3–20). New York: Guilford Press.

Leff, M. (1980). Interpretation and the art of the rhetorical critic. *The Western Journal of Speech Communication, 44,* 337–349.

Leopold, A. (1987). *A Sand County almanac.* Oxford, England: Oxford Press. (Reprint of original 1949 edition).

Leroux, N. R. (1992). Perceiving rhetorical style: Toward a framework for criticism. *Rhetoric Society Quarterly, 22,* 29–44.

Lewicki, R. J., Saunders, D. M., & Minton, J. W. (1999). *Negotiation* (3rd ed.). Boston: Irwin/McGraw-Hill.

Lewicki, R. J., & Wiethoff, C. (2000). Trust, trust development, and trust repair. In M. Deutsch & P. Coleman (Eds.), *Handbook of conflict resolution* (pp. 90–95). San Francisco: Jossey-Bass.

Linenthal, E. T. (1995). *Preserving memory: The struggle to create America's Holocaust museum.* New York: Penguin.

Little, J. B. (1996, May 13). Consensus even came to Washington, D.C. *High Country News, 28*(9). [On-line]. Available: http://www.hcn.org

Lugones, M. (1994). Playfulness, "world"-travelling, and loving perception. In D. S. Madison (Ed.), *The woman that I am: The literature and culture of contemporary women of color* (pp. 560–573). New York: St. Martin's Press.

Lynn, F. M., & Busenberg, G. J. (1995). Citizen advisory committees and environmental policy: What we know, what's left to discover. *Risk Analysis, 15,* 147–162.

Lyon, L. (1987). *The community in urban society.* Chicago: Dorsey Press.

Lyotard, J-F. (1984). *The postmodern condition: A report on knowledge.* Minneapolis: University of Minnesota Press.

MacCannell, D. (1976/1999). *The tourist: A new theory of the leisure class.* Berkeley: University of California Press.

Macedo, S. (Ed.). (1999). *Deliberative politics: Essays on democracy and disagreement.* New York: Oxford University Press.

Magill, A. W. (1988). Natural resource professionals: The reluctant public servants. *The Environmental Professional, 10*(4), 295–303.

Magill, A. W. (1991). Barriers to effective public interaction: Helping natural resource professionals adjust their attitudes. *Journal of Forestry, 89*(10), 16–18.

Mapping what's happening on the ground: Three toxic tour surveys. (2001, January–March). *EJ Times: The Sierra Club environmental justice newsletter ii*(1), 7–10. [On-line]. Available: http://www.sierraclub.org/environmental_justice/newsletter/v2i1.pdf

Marston, B. (2001, May 7). A modest chief moved the Forest Service miles down the road. *High Country News, 33*(9). [On-line]. Available: http://www.hcn.org

Mazmanian, D. A., & Nienaber, J. (1979). *Can organizations change?* Washington, DC: Brookings Institute.

McCarthy, T. (1989). Introduction. T. Burger (Trans.). In J. Habermas (Ed.), *The structural transformation of the public sphere* (pp. xi–xiv). Cambridge, MA: MIT.

McCarthy, T. (1992). Practical discourse: On the relation of morality to politics. In C. Calhoun (Ed.), *Habermas and the public sphere* (pp. 51–72). Cambridge: MIT.

McComas, K. A. (2001). Theory and practice of public meetings. *Communication Theory, 11*(1), 36–55.

McCracken, G. (1988). *The long interview.* Newbury Park, CA: Sage.

McGee, M. C. (1980/1995). The "ideograph": A link between rhetoric and ideology. In C. R. Burchardt (Ed.), *Readings in rhetorical criticism* (pp. 442–456). State College, PA: Strata.

McLain, A. W. (1995, August). *Who decides? Policy processes, federalism, and ecosystem management.* Concept paper for the Eastside Assessment Team. Seattle: University of Washington, College of Forest Resources. [On-line]. Available: http://www.icbemp.gov/science/mclain.pdf

Melcer, R. (1999). CDC rejects Fernald study. *Cincinnati Enquirer.* [On-line]. Available: http://enquirer.com/editions/1999/09/24/loc_cdc_rejects_fernald.html

Michelman, F. I. (1997). How can the people ever make the laws? A critique of deliberative democracy. In J. Bohman & W. Rehg (Eds.), *Deliberative democracy* (pp. 145–172). Cambridge, MA: MIT University Press.

Miller, B. (2000). Media art and activism: A model for collective action. In R. Hofrichter (Ed.), *Reclaiming the environmental debate: The politics of health in a toxic culture* (pp. 313–325). Cambridge, MA: MIT University Press.

Moore, M. P. (1993). Constructing irreconcilable conflict: The function of synecdoche in the spotted owl controversy. *Communication Monographs, 60,* 258–274.

Moote, M. A., Brown, B. A., Kingsley, E., Lee, S. X., Marshall, S., Voth, D. E., & Walker, G. B. (2001). Process: Redefining relationships. In G. J. Gray, M. J. Enzer, & J. Kusel (Eds.), *Understanding community-based forest ecosystem management* (pp. 97–116). New York: Food Products Press.

Mouffe, C. (1999). Deliberative democracy or agonistic pluralism. *Social Research, 66,* 745–758.

Mowrey, M., & Redmond, R. (1993). *Not in our backyard.* New York: Wm. Morrow.

Muir, S. A., & Veenendall, T. L. (Eds.). (1996). *Earthtalk: Communication empowerment for environmental action.* Westport, CT: Praeger.

Mumpower, J. (1995). The Dutch study groups revisited. In O. Renn, T. Webler, & P. Wiedemann (Eds.), *Fairness and competence in citizen participation: Evaluating models for environmental discourse* (pp. 321–337). Boston: Kluwer.

Murdock, B. S., & Sexton, K. (1999). Community based partnerships. In K. Sexton, A. A. Marcus, K. W. Easter, & T. D. Burkhardt (Eds.), *Better environmental decisions: Strategies for governments, businesses and communities* (pp. 377–400). Washington, DC: Island Press.

Murphy, J. M. (1994, Summer). Presence, analogy, and Earth in the balance. *Argumentation and Advocacy, 31,* 1–16.

Murray, J., & Nobleza, M. (2001, March 8–9). *Building trust: Lessons from the Nile River Basin.* Conference on Evaluating Environmental and Public Policy Dispute Resolution Programs and Policies. Washington, DC: Alan K. Campbell Public Affairs Institute.

National Environmental Justice Advisory Council. (2000). *The model plan for public participation.* U.S. Environmental Protection Agency, EPA-300-K-00-001.

National Environmental Policy Act Handbook. (1988). BLM Manual. Rel. 1-1547.

National Environmental Policy Act (NEPA) of 1970, S. 40, 416, 115th Congress. (1969).

Nayyar, D. (2000, August). *Introduction: The new role and functions for the UN and Bretton Woods institutions.* Helsinki, Finland: The United Nations University and the World Institute for Development Economics Research. [On-line]. Available: http://www.wider.unu.edu/publications/1998-1999-5-1/1998-1999-5-1.pdf

Neumann, M. (1999). *On the rim: Looking for the grand canyon.* Minneapolis: University of Minnesota.

New York State Environmental Conservation Law. (1995). *McKinney's Consolidated Laws of New York,* Section 8-019. St. Paul, MN: West Group.

New York Times. (1969a, August 17). p. 61. [On-line]. Available: Lexis Nexus.

New York Times. (1969b, Nov. 8). p. 18. [On-line]. Available: Lexis Nexus.

New York Times. (1969c, Dec. 3). p. 56. [On-line]. Available: Lexis Nexus.

North American Free Trade Agreement. (1994). [On-line]. Available: http://www.ftaa-alca.org/alca_e.asp

Novotny, P. (1998). Popular epidemiology and the struggle for community health in the environmental justice movement. In D. Faber (Ed.), *The struggle for ecological democracy: Environmental justice movements in the United States* (pp. 137–158). New York: Guilford.

NRC. (1994). *Building consensus through risk assessment and management of the Department of Energy's environmental remediation program.* Washington, DC: National Academy Press.

NRC. (1997). *A review of the Radiological Assessment Corporation's Fernald dose reconstruction report.* Washington, DC: National Academy Press.

Nuttal, M. (1997). Packaging the wild: Tourism development in Alaska. In S. Abram, J. Waldren, & D. V. L. MacLeod (Eds.), *Tourists and tourism: Identifying with people and places* (pp. 223–238). Oxford, England: Berg.

O'Brien, M. H. (2000). When harm is not necessary. In R. Hofrichter (Ed.), *Reclaiming the environmental debate: The politics of health in a toxic culture* (pp.113–134). Cambridge, MA: MIT Press.

OD Corp. (1996). *The Seventh American Forest Congress.* February 20–24, 1996. [On-line]. Available: http://www.tmn.com/od/forestconf.htm

Office of U.S. Trade Representative. (1998, July 28). Request for public comment on the Committee of Government Representatives on the participation of civil society in connection with the Free Trade Area of the Americas. 63 Fed. Reg. 40379.

Olwig, K. R. (1996). Reinventing common nature: Yosemite and Mount Rushmore—A meandering tale of a double nature. In W. Cronon (Ed.), *Uncommon ground: Rethinking the human place in nature* (pp. 379–408). New York: W. W. Norton.

O'Neal, J. (2000). For generations yet to come: Junebug Productions' environmental justice project. In R. Hofrichter (Ed.), *Reclaiming the environmental debate: The politics of health in a toxic culture* (pp. 313–325). Cambridge, MA: MIT University Press.

Opie, J., & Elliott, N. (1996). Tracking the elusive jeremiad: The rhetorical character of American environmental discourse. In J. Cantrill & C. L. Oravec (Eds.), *The symbolic earth: Discourse and our creation of the environment* (pp. 9–37). Lexington: University Press of Kentucky.

Oravec, C. (1981). John Muir, Yosemite, and the sublime response: A study in the rhetoric of preservationism. *Quarterly Journal of Speech, 67*(3), 245–258.

Oravec, C. (1984). Conservationism vs. preservationism: The "public interest" in the Hetch-Hetchy controversy. *Quarterly Journal of Speech, 70,* 444–458.

Oravec, C. L. (1996). To stand outside oneself: The sublime in the discourse of natural scenery. In J. Cantrill & C. L. Oravec (Eds.), *The symbolic earth: Discourse and our creation of the environment* (pp. 58–75). Lexington: University Press of Kentucky.

Organization of Economic Cooperation and Development. (1960, December 14). *Convention on the Organisation for Economic Co-operation and Development.* Paris: OECD.

Osborne, D., & Gaebler, T. (1992). *Reinventing government.* Reading, MA: Addison-Wesley.

Owens, S. (2000). "Engaging the public": Information and deliberation in environmental policy. *Environment and Planning A, 32,* 1141–1148.

Palmer, D. (2001, March 7). *U.S. trade representative sued for hiding documents* [wire service report]. Washington: Reuters. [On-line]. Available: http://www.globalexchange.org/ftaa/news2001/reuters030701.html.

Parks, G. (1965/1986). *A choice of weapons.* St. Paul: Minnesota Historical Society Press.

Pearce, W. B. (1995). A sailing guide for social constructionists. In W. Leeds-Hurwitz (Ed.), *Social approaches to communication* (pp. 88–113). New York: Guilford Press.

Pearce, W. B., & Littlejohn, S. W. (1997). *Moral conflict: When social worlds collide*. Newbury Park, CA: Sage.

Peele, E. (1990, August 28). Citizen advisory groups: Improving their effectiveness. *Proceedings of the 12th Annual Department of Energy Low-Level Waste Management Conference,* Chicago, IL.

Pennock, J. R., & Chapman, J. W. (1975). *Participation in politics: Nomos XVI.* New York: Lieber-Atherton.

Perelman, C. H., & Olbrechts-Tyteca, L. (1969). *The new rhetoric: A treatise on argumentation.* (J. Wilkinson, & P. Weaver, Trans.). Notre Dame, IN: University of Notre Dame Press. (Original work published 1958).

Persons, G. A. (1990, March). Defining the public interest: Citizen participation in metropolitan and state policy making. *National Civic Review, 79,* 118–131.

Peterson, T. R. (1991). Telling the farmers' story: Competing responses to soil conservation rhetoric. *Quarterly Journal of Speech, 77,* 289–308.

Peterson, T. R. (1997). *Sharing the earth: The rhetoric of sustainable development.* Columbia: University of South Carolina Press.

Peterson, T. R. (1998). Environmental communication: Tales of life on earth. *Quarterly Journal of Speech, 84*(3), 371–393.

Peterson, T. R., & Horton, C. C. (1995). Rooted in the soil: How understanding the perspectives of landowners can enhance the management of environmental disputes. *Quarterly Journal of Speech, 81*(2), 139–166.

Peterson, T. R., & Peterson, M. J. (1996). Valuation analysis in environmental policy making: How economic models limit possibilities for environmental advocacy. In J. G. Cantrill, & C. L. Oravec (Eds.), *The symbolic earth: Discourse and our creation of the environment* (pp. 198–218). Lexington: University Press of Kentucky.

Peterson, T. R., Witte, K., Enkerlin-Hoeflich, E., Espericueta, L., Flora, J. T., Florey, N., Loughran, T., & Stuart, R. (1994). Using informant-directed interviews to discover risk orientation: How formative evaluations based in interpretative analysis can improve persuasive safety campaigns. *Journal of Applied Communication Research, 22,* 199–215.

Pezzullo, P. C. (2001, Winter). Performing critical interruptions: Stories, rhetorical invention, and the environmental justice movement. *Western Journal of Communication, 65*(1), 1–25.

Pfleger, K. (2001). Bush to revise forest road-ban rules. *Newsday.com.* [Online]. Available: http://www.newsday.com

Phillips, N., & Higgott, R. (1999, November). Global governance and the public domain: Collective goods in a 'Post-Washington consensus' Era. *Working Paper No.47/49*. Coventry, England: University of Warwick Center for the Study of Globalisation and Regionalisation.

Pieterse, J. N. (2000). Shaping globalization. In J. N. Pieterse (Ed.), *Global futures: Shaping globalization* (pp. 1–19). London: Zed Books.

Plant, J., & Plant, C. (1992). *Putting power in its place: Create community control!* Philadelphia: New Society Publishers.

Platt, K. (1997). Chicana struggles for success and survival: Cultural poetics of environmental justice from the Mothers of East Los Angeles. *Frontiers, 18,* 48–63.

Plough, A., & Krimsky, S. (1987). The emergence of risk communication studies: Social and political context. *Science, Technology, and Human Values, 12*(3–4), 4–10.

Popovic, N. A. F. (1993, Spring). The right to participate in decisions that affect the environment. *Pace Environmental Law Review, 10*(2), 683–709.

Proctor, R. N. (1991). *Value-free science? Purity or power in modern knowledge.* Cambridge, MA: Harvard University Press.

Public Citizen. (1998, August 21). *Comments of Public Citizen on the Office of the United States Trade Representative's request for public comment on the Committee of Government Representatives on the Participation of Civil Society in connection with the Free Trade Area of the Americas negotiations (Comments submitted to the Office of the USTR).* Washington, DC: Author.

Public Law 88-577. (1964, September 3). *Wilderness Act.*

Public Law 94-588. (1976, October 22). *National Forest Management Act.*

Public Law 95-495. (1978, October 21). *The Boundary Waters Act.*

Pugh, C. (1997, October 19). A force of one. *TEXAS-Houston Chronicle Magazine,* pp. 8–14.

Putnam, R. (1993). *Making democracy work: Civic traditions in modern Italy.* Princeton, CT: Princeton University Press.

Putnam, R. D. (1995, December). Tuning in, tuning out: The strange disappearance of social capital in America. *Political Science & Politics, 28,* 664–683.

Quinn, R. E., & Rohrbaugh, J. W. (1981). A competing values approach to organizational analysis. *Public Productivity Review, 5,* 141–159.

Quinn, R. E., & Rohrbaugh, J. W. (1983). A spatial model of effectiveness criteria: Towards a competing values approach to organizational effectiveness. *Management Science, 29*(1), 363–377.

Rabe, B. C. (1991). Impediments to environmental dispute resolution in the American political context. In M. K. Mills (Ed.), *Alternative dispute resolution in the private sector* (pp. 143–163). Chicago, IL: Nelson Hall.

RAC. (1998). *Task 6: Radiation doses and risk to residents from FMPC operations from 1951–1988.* Volume II (Final Report). Neeses, SC: RAC.

Reagan, P., & Rohrbaugh, J. (1990). Group decision process effectiveness: A competing values approach. *Group & Organization Studies, 15*(1), 20–43.

Reason, P. (1994). Three approaches to participatory inquiry. In N. K. Denzin & Y. S. Lincoln (Eds.), *Handbook of qualitative research* (pp. 324–339). Newbury Park, CA: Sage.

Regulations for implementing the procedural provisions of the National Environmental Policy Act. (1978). Council on Environmental Quality, Office of the President. *Federal Register, 43,* 55978–56007.

Reich, M. R. (1991). *Toxic politics: Responding to chemical disasters.* Ithaca, NY: Cornell University Press.

Reich, R. B. (Ed.). (1988). *The power of public ideas.* Cambridge, MA: Ballinger.

Reinecke, W. H. (1998). *Global public policy: Governing without government?* Washington, DC: Brookings Institution Press.

Renn, O. (1992). Concepts of risk: A classification. In S. Krimsky & D. Golding (Eds.), *Social theories of risk* (pp. 53–82). Westport, CT: Praeger.

Renn, O., Webler, T., & Wiedemann, P. (Eds.). (1995). *Fairness and competence in citizen participation: Evaluating models for environmental discourse.* Boston: Kluwer Academic Press.

Richardson, H. S. (1997). Democratic institutions. In J. Bohman, & W. Rehg (Eds.), *Deliberative democracy* (pp. 349–382). Cambridge, MA: MIT Press.

Roadless Area Conservation (2000). *Participation summary. USDA-Forest Service.* [On-line]. Available: http://www.roadless.fs.fed.us/documents

Roadless Area Conservation (2001a). *PowerPoint roadless presentation. USDA-Forest Service.* [On-line]. Available: http://www.roadless.fs.fed.us/documents/powerpt/index

Roadless Area Conservation (2001b). *Public comment. USDA-Forest Service.* [On-line]. Available: http://www.roadless.fs.fed.us/publiccomment/

Roberts, P. (1996). Habermas, philosophies, and Puritans: Rationality and exclusion in the dialectical public sphere. *Rhetoric Society Quarterly, 26,* 47–65.

Rodrik, D. (1997). *Has globalization gone too far?* Washington, DC: Institute for International Economics.

Roosevelt, T. R. (1913/1920). *Theodore Roosevelt, an autobiography.* New York: Charles Scribner's Sons.

Rosenbaum, N. M. (1978). Citizen participation and democratic theory. In S. Langton (Ed.), *Citizen participation in America: Essays on the state of the art* (pp. 43–54). Lexington, MA: Lexington Books.

Rowan, K. E. (1994). The technical and democratic approaches to risk situations: Their appeal, limitations, and rhetorical alternative. *Argumentation, 8,* 391–409.

Rowe, G., & Frewer, L. J. (2000). Public participation methods: A framework for evaluation. *Science, Technology, and Human Values, 25*(1), 3–29.

Russell, P., & Hines, B. (1992). *Savannah: A history of her people since 1733.* Savannah, GA: Fredrick C. Beil.
Sachs, J. (1997, December 11). IMF is a power unto itself. *Financial Times*, p. 21.
Sandel, M. J. (1996, March). America's search for a new public philosophy. *The Atlantic Monthly,* pp. 57–74.
Sandmann, W. (1996). The rhetorical function of *Earth in the balance*. In S. A. Muir, & T. L. Veenendall (Eds.), *Earthtalk: Communication and empowerment for environmental action* (pp. 119–134). Westport, CT: Praeger.
Sarason, S. B. (1974). *The psychological sense of community: Prospects for a community psychology.* San Francisco: Jossey-Bass.
Sarno, D. J. (2001, Third Quarter). The future of Fernald process: Creating a community vision and legacy. *IAP2 News: Improving the Practice,* 1–8.
Schell, J. (1982). *The fate of the Earth*. New York: Avon.
Schmitt, B., & Guidera, A. (1998, October 20). Ports deepening to be put on hold. *The Savannah Morning News Electronic Edition.* [On-line]. Available: http://www.savannahnow.com
Scholte, J. A. (2000). *Globalization: A critical introduction.* New York: St. Martin's Press.
Scholte, J. A. (1998, May). *Civil society and the International Monetary Fund: An underdeveloped dialogue.* Institute of Social Studies, The Hague. Paper for Discussions in Washington, DC.
Schutz, W. G. (1958*). FIRO: A three dimensional theory of interpersonal behavior.* New York: Holt, Rinehart and Winston.
Schwerin, E. W. (1995). *Mediation, citizen empowerment and transformational politics.* Westport, CT: Praeger.
Sclove, R. E. (1995). *Democracy and technology.* New York: Guilford.
Scott, W. R. (1995). *Institutions and organizations.* Thousand Oaks, CA: Sage.
Seabrook, C. (1998, November 22). A not so safe harbor: A proposal to deepen Savannah's port could imperil the wetlands of the wildlife refuge. *The Atlanta Journal Constitution,* p. 1B.
Sechler, B. (1998, August 8). Harbor deepening gets county support. *The Savannah Morning News Electronic Edition.* [On-line]. Available: http://www.savannahnow.com
SEG. (1999, February 6). Meeting tapes. Savannah, Georgia: Georgia Ports Authority.
SELC. (2000). *Savannah River, SC and GA.* [On-line]. Available: http://www.selc.org
Senge, P. (1990). *The fifth discipline: The art & practice of the learning organization.* New York: Currency Doubleday.
Setterberg, F., & Shavelson, L. (1993). *Toxic nation: The fight to save our communities from chemical contamination.* New York: John Wiley & Sons.

Sewell, W. R. D., & Phillips, S. D. (1979). Models for the evaluation of public participation programmes. *Natural Resources Journal, 19*(2), 337–358.

Shabecoff, P. (1993). *A fierce green fire: The American environmental movement.* New York: Hill & Wang.

Shamsie, Y. (2000, January). *Engaging with civil society: Lessons from the OAS, FTAA, and Summit of the Americas. Ottawa, Canada.* (Available from the North-South Institute/L'Institut Nord-Sud, 55 Murray, Suite 200, Ottawa, Canada KIN 5M3.)

Shannon, M. A. (1991, June). *Building public decisions, learning through planning: An evaluation of the NFMA forest planning process.* Report for the Office of Technology Assessment's Evaluation of the Renewable Resources Planning Technologies for the National Forests.

Shannon, M. A. (1992). Community governance: An enduring institution of democracy. In U.S. Senate Report. *Symposium on multiple use and sustained yield: Changing philosophies for federal land management* (pp. 219–250). Washington, DC: Congressional Research Service, Library of Congress.

Shapiro, I. (1999). Enough of deliberation: Politics is about interest and power. In S. Macedo (Ed.), *Deliberative politics: Essays on democracy and disagreement* (pp. 43–62). New York: Oxford University Press.

Shepherd, A., & Bowler, C. (1997). Beyond the requirements: Improving public participation in EIA. *Journal of Environmental Planning and Management, 40*(6), 725–738.

Shindler, B., & Neburka, J. (1997). Public participation in forest planning: Eight attributes of success. *Journal of Forestry, 95*(1), 17–19.

Shindler,B., Steel, B., & List, P. (1996, June). Public judgments of adaptive management: A response from forest communities. *Journal of Forestry, 94*(6), 4–12.

Short, B. (1991). Earth First! and the rhetoric of moral confrontation. *Communication Studies, 42*(2), 172–188.

Shotter, J. (1993). *Conversational realities: Constructing life through language.* London: Sage Publications.

Sieg, E. C. (1985). *Eden on the marsh: An illustrated history of Savannah.* Northridge, CA: Windsor Publications.

Simmons, P. J. (1998, Fall). Learning to live with NGOs. *Foreign Policy.* [On-line]. Available: http://www.globalpolicy.org/ngos/issues/simmons.htm

Sirianni, C., & Friedland, L. (1995). *Civic environmentalism.* [On-line]. Available: http://www.cpn.org/sections/topics/env_spectives/civic_environmentalismA.html

Slovic, P. (1987). Perception of risk. *Science, 236,* 280–285.

Slovic, S. (1996). Epistemology and politics in American nature writing: Embedded rhetoric and discrete rhetoric. In C. G. Herndl & S. C. Brown (Eds.), *Green culture: Environmental rhetoric in contemporary America* (pp. 82–110). Madison: University of Wisconsin Press.

Smith, D. (1999, June 22). *Greenpeace responds to Louisiana Governor Mike Foster.* [On-Line]. Available: http://www.greepeaceusa.org/toxics/fosterltertext.htm

Smith, R. (1997, July). M.O.S.E.S. leads Winona, Texas, to environmental justice. *The New Crisis, 104,* 30–32.

Snow, D. A., & Benford, R. D. (1992). Master frames and cycles of protest. In A. D. Morris, & C. M. Mueller (Eds.), *Frontiers in social movement theory* (pp. 133–155). London: Yale University Press.

Spears, T. (1989, May 3). Tours to focus on pollution in the city. *The Toronto Star,* p. B7.

Spyke, N. P. (1999). Public participation in environmental decisionmaking at the new millennium: Structuring new spheres of public influence. *Boston College Environmental Affairs Law Review, 26*(2), 263–313.

Steelman, T. A., & Ascher, W. (1997). Public involvement methods in natural resource policy making: Advantages, disadvantages, and trade-offs. *Policy Sciences, 30*(2), 71–90.

Stephenson, A. M. (2000, March 25). *The state of the FTAA negotiations at the turn of the millennium.* Paper presented at the conference on Trade and the Western Hemisphere, Southern Methodist University, Dallas, TX.

Stephenson, M. O., & Pops, G. M. (1991). Public administration and conflict resolution: Democratic theory, administrative capacity, and the case of negotiated rule-making. In M. K. Mills (Ed.), *Alternative dispute resolution in the public sector* (pp. 13–26). Chicago: Nelson Hall.

Stewart, J. (1995a). *Language as articulate contact: Toward a post-semiotic philosophy of communication.* Albany: State University of New York.

Stewart, J. (1995b). Philosophical features of social approaches to interpersonal communication. In W. Leeds-Hurwitz (Ed.), *Social approaches to communication* (pp. 23–45). New York: Guilford Press.

Stiglitz, J. (1998, October 19). Towards a new paradigm for development: Strategies, policies and processes. *The 1998 Prebisch Lecture at UNCTAD.* Geneva, Switzerland.

Swerczek, M. (1999, February 20). Fifty protest Shell plant near Norco. *The Times-Picayune,* p. B3.

Taxpayers for Common Sense. (1999). *Savannah Harbor expansion: Traffic forecasts wildly optimistic* [On-line]. Available: http://www.taxpayer.net

Taylor, B. C. (1998). Nuclear weapons and communication studies: A review essay. *Western Journal of Communication, 54,* 395–419.

Taylor, C. A. (1996). *Defining science: A rhetoric of demarcation.* Madison: University of Wisconsin Press.

Tieleman, K. (2000). *The failure of the Multilateral Agreement on Investment (MAI) and the absence of a global public policy network. Case study for the U.N. Vision Project on global public policy networks.* [On-line]. Available: http://www.globalpublicpolicy.net/Tieleman%20GPP%202000.pdf

Tipple, T. J., & Wellman, J. D. (1989). Life in the fishbowl: Public participation rewrites public foresters' job descriptions. *Journal of Forestry, 87*(3), 24–30.

Toumey, C. P. (1996). *Conjuring science: Scientific symbols and cultural meanings in American life.* New Brunswick, NJ: Rutgers University Press.

Trumbo, C. W. (2000). Reasons why: Describing polarized orientations toward perceptions of environmental risk. In N. W. Coppola, & B. Karis (Eds.), *Technical communication, deliberative rhetoric, and environmental discourse: Connections and directions* (pp. 191–224). Stamford, CT: Ablex.

Tuler, S., & Webler, T. (1995). Process evaluation for discursive decision making in environmental and risk policy. *Human Ecology Review 2*(1), 62–71.

Tuler, S., & Webler, T. (1999). Voices from the forest: What participants expect of a public participation process. *Society & Natural Resources, 12,* 437–453.

Twarkins, M., Fisher, L., & Robertson, T. (2001). Public involvement in forest management planning: A view from the Northeast. *Journal of Sustainable Forestry, 13*(1–2), 237–251.

Twight, B. W. (1977). Confidence or more controversy: Whither public involvement? *Journal of Forestry, 75*(2), 93–95.

Ulman, H. L. (1996). "Thinking like a mountain": Persona, ethos, and judgment in American nature writing. In C. G. Herndl & S. C. Brown (Eds.), *Green culture: Environmental rhetoric in contemporary America* (pp. 46–81). Madison: University of Wisconsin Press.

United Church of Christ Commission for Racial Justice. (1987). In B. A. Goldman & L. Fitton (Eds.), *Toxic wastes & race in the United States.* New York: UCC.

US BLM. (1998). *Environmental assessment for the Western Mobile proposed gravel-pit site* (BLM Publication No. 5653–5610). Albuquerque, NM: Author.

U.S. Congress. (1995). *Oversight hearings, Voyageurs National Park and Boundary Waters Canoe Area.* Serial No. 104-43.

USDA-FS. 1981. *Plan to implement the Boundary Waters Canoe Area Wilderness Act.*

USDA-FS. (1986a). *Land and resource management plan, Superior National Forest.* Washington, DC: U.S. Government Printing Office.

USDA-FS. (1986b). *Record of decision: Final environmental impact statement and land and resource management plan, Superior National Forest.* Washington, DC: U.S. Government Printing Office.

USDA-FS. (1991, January). *Improving communications and working relationships.* Final report of the Committee on External Communications. Washington, DC: U.S. Government Printing Office.

USDA-FS. (1992). *Public participation handbook.* Washington, DC: U.S. Government Printing Office.

USDA-FS. (1993a). *BWCA wilderness management plan and implementation schedule, Superior National Forest.* Washington, DC: U.S. Government Printing Office.

USDA-FS. (1993b). *Record of Decision: Final environmental impact statement for the BWCA wilderness management plan and implementation schedule, Superior National Forest.* Washington, DC: U.S. Government Printing Office.

USDA-FS, & USDOI-BLM. (1994). *Record of decision for amendments to Forest Service and Bureau of Land Management planning documents within the range of the northern spotted owl and Standards and guidelines for management of habitat for late-successional and old-growth forest related species within the range of the northern spotted owl.* Washington, DC: U.S. Government Printing Office, 1994-589-111/00001 Region No. 10.

US DOE. (1991, November). *Public participation in environmental restoration activities.* Washington, DC: US DOE, Office of Environmental Guidance RCRA/CERCLA Division (EH-231).

US DOE. (1995a). *Environmental management public participation.* Office of Environmental Management. [On-line]. Available: http://www.em.doe.gov/public/empubpar.html

US DOE. (1995b, January). *Closing the circle on the splitting of the atom: The environmental legacy of nuclear weapons production in the United States and what the Department of Energy is doing about it.* Washington, DC: DOE, Office of Environmental Management.

US DOE. (1997, January). *Linking legacies: Connecting the Cold War nuclear weapons production processes to their environmental consequences.* Washington, DC: DOE, Office of Environmental Management.

US DOE-FEMP. (1994, October). *Fact sheet.* Fernald, OH: DOE-FEMP.

US DOE-FEMP. (1995, January). *Community relations plan.* Fernald, OH: DOE-FEMP.

US DOE-FN. (1993, November). *Public involvement program for the Fernald Environmental Management Project.* Fernald, OH: DOE-FEMP.

US EPA. (1997a). *Annual report on reinvention.* [On-line]. Available: http://www.epa.gov

US EPA. (1997b). *Managing for better environmental results.* [On-line]. Available: http://www.epa.gov

US EPA. (n.d.). *Report of the environmental justice and compliance assurance roundtable, Oct. 17–19, 1996.* San Antonio, TX: National Environmental Justice Advisory Council. [On-line]. Available: http://es.epa.gov/oeca/main/ej/nejac/pdf/1096.pdf

US FWS. (2000). *National Environmental Policy Act reference handbook.* [Online]. Available: http://www.fws.gov/r9esnepa/

US GAO. (1983). *Siting of hazardous waste landfills & their correlation with racial and economic status of surrounding communities.* Washington, DC: U.S. Government Printing Office.

U.S. Senate. (1996). *Hearing on S. 1738, Boundary Waters Canoe Area Wilderness Accessibility and Partnership Act.* Senate Hearing 104-717.

Vari, A. (1995). Citizens' advisory committee as a model for public participation: A multiple-criteria evaluation. In O. Renn, T. Webler, & P. Wiedemann (Eds.), *Fairness and competence in citizen participation: Evaluating models for environmental discourse* (pp. 103–115). Boston: Kluwer.

Vaughan, E., & Seifert, M. (1992). Variability in the framing of risk issues. *Journal of Social Issues, 48*(4), 119–135.

Waddell, C. (1996). Saving the Great Lakes: Public participation in environmental policy. In C. G. Herndl & S. C. Brown (Eds.), *Green culture: Environmental rhetoric in contemporary America* (pp. 141–165). Madison: University of Wisconsin Press.

Walker, G. B., & Daniels, S. E. (1997). Collaborative public participation in environmental conflict management: An introduction to five approaches. In S. L. Senecah (Ed.), *Proceedings of the Fourth Biennial Conference on Communication and Environment* (pp. 271–289). Syracuse: State University of New York College of Environmental Science and Forestry.

Walker, G. B., & Daniels, S. E. (2001). Natural resource policy and the paradox of public involvement: Bringing scientists and citizens together. *Journal of Sustainable Forestry, 13*(1–2), 253–269.

Walker, G. B., Daniels, S. E., Blatner, K. A., & Carroll, M. S. (1996). *Civic discovery and ecosystem-based management: Collaborative learning in fire recovery planning.* Paper presented at the Speech Communication Association Convention, San Diego, CA.

Wallach, L. (1993). Hidden dangers of GATT and NAFTA. In R. Nader, W. Greider, M. Atwood, D. Philips, & P. Choate. (Eds.), *The case against "free trade": GATT, NAFTA, and the globalization of corporate power* (pp. 23–64). San Francisco: Earth Island.

Watts, M. (2000). Poverty and the politics of alternatives at the end of the millenium. In J. N. Pieterse (Ed.), *Global futures: Shaping globalization* (pp. 133–147). London: Zed Books.

Weaver, B. J. (1996). "What to do with the mountain people?" The darker side of the successful campaign to establish the Great Smoky Mountains National Park. In J. G. Cantrill & C. L. Oravec (Eds.), *The symbolic earth: Discourse and our creation of the environment* (pp. 151–175). Lexington: University Press of Kentucky.

Weber, E. P. (1998). *Pluralism by the rules: Conflict and cooperation in environmental regulation.* Washington, DC: Georgetown University Press.

Webler, T. (1995). "Right" discourse in citizen participation: An evaluative yardstick. In O. Renn, T. Webler, & P. Wiedemann (Eds.), *Fairness and competence in citizen participation: Evaluating models for environmental discourse* (pp. 35–86). Boston: Kluwer.

Webler, T., Tuler, S., & Krueger, J. R. (2001). What is a good public participation process? Five perspectives from the public. *Environmental Management 27*(3), 435–450.

Wehr, P. (1979). *Conflict regulation.* Boulder, CO: Westview.

Weisbrot, M., Naiman, R., & Kim, J. (2000). *The emperor has no growth: Declining economic growth rates in the era of globalization.* Washington, DC: Center for Economic and Policy Research. [On-line]. Available: http://www.cepr.net/IMF/The_Emperor_Has_No_Growth.htm

Wellman, J. D., & Tipple, T. J. (1990). Public forestry and direct democracy. *The Environmental Professional, 12*(1), 77–86.

Wengert, N. (1976). Citizen participation: Practice in search of a theory. *Natural Resource Journal, 16*(1), 23–40.

Whyte, W. F. (1984). *Learning from the field: A guide from experience.* Beverly Hills, CA: Sage.

Wieboldt, B. (1995, March 14). Presentation at Land-Use Seminar sponsored by the New York State Forum for Conflict and Consensus, Albany, NY.

Williams, B. A., & Matheny, A. R. (1995). *Democracy, dialogue, and environmental disputes: The contested languages of social regulation.* New Haven, CT: Yale University Press.

Willis, D. (1998, February 2). Medical incinerator is causing environmental harm, group says; 'Toxic Tour' travels to area industrial sites. *The Baltimore Sun,* p. 3B.

Wondolleck, J. M., & Yaffee, S. L. (2000). *Making collaboration work: Lessons from innovation in natural resource management.* Washington, DC: Island Press.

Wood, A. (2001, January). *Options for reforming the IMF's governance structure.* [On-line]. Available: http://www.brettonwoodsproject.org/topic/reform/sapimf.html

World Bank. (1994, March). *The World Bank policy on disclosure of information.* Washington, DC: Author.

Yaffee, S. L. (1994). *The wisdom of the spotted owl: Policy lessons for a new century.* Washington, DC: Island Press.

Yaffee, S. L. (1996). Ecosystem management in practice: The importance of human institutions. *Ecological Applications, 6*(3), 724–727.

Yankelovich, D. (1981). *New rules.* New York: Random House.

Yosie, T. F., & Herbst, T. D. (1998). *Using stakeholder processes in environmental decision making: An evaluation of lessons learned, key issues, and future challenges.* [On-line]. Retrieved April 10, 1999. Available: http://www.riskworld.com/Nreports/1998stakehold/html

Zuckerman, M. (1998). A second american century. *Foreign Affairs, 77*(3), 18–31.

CONTRIBUTORS

J. ROBERT COX (Ph.D., University of Pittsburgh) is a professor in the Department of Communication Studies and in the Ecology Curriculum at the University of North Carolina at Chapel Hill. From 1994 to 1996, and from 2000 to 2001, Cox served as the president of the national Sierra Club. His research in rhetorical theory has examined different rhetorical tensions and modes of advocacy in the civil rights, child labor, antiwar, and environmental movements as well as in organized labor. His recent work has focused on the barriers to political voice in low-income communities and in communities of color, particularly as these face challenges from environmental threats.

JOHN W. DELICATH formerly an assistant professor and research associate in the Center for Environmental Communication Studies at the University of Cincinnati, works as a communications analyst with the Natural Resources and Environment group at the U.S. General Accounting Office. He is also an independent scholar writing about brownfields, environmental justice, cultural activism, and public participation in environmental decision making. He was co-planner of the Sixth Biennial Conference on Communication & Environment, and recently completed work with local and state organizations to produce a citizens' guide to brownfields in southwest Ohio. He also worked with the Miami Group of the Sierra Club and local environmental justice groups to produce the video "The Faces of Environmental Injustice in Cincinnati" (Sunshine Productions).

STEPHEN P. DEPOE (Ph.D., Northwestern University) is an associate professor and director of the Center for Environmental Communication Studies (www.uc.edu/cecs). During the past decade, he has worked on projects and published essays related to public participation and environmental decision making in the greater Cincinnati area, including the Fernald Environmental Management Project. In 2001, he served as co-planner of the Sixth Biennial Conference on Communication and Environment.

AMANDA C. GRAHAM (Ph.D., University of Washington) is currently the Education Program Manager at the Laboratory for Energy and the Environment at the Massachusetts Institute of Technology. Her professional and research interests center on the theory and practice of dialogue in environmental education and decision making. Her doctoral dissertation developed a tensional approach to civic environmental deliberation that interprets oppositions as interdependent and mutually transformative, and applied this framework to the Pacific salmon controversy in the northwestern United States.

JENNIFER DUFFIELD HAMILTON (Ph.D. candidate, University of Cincinnati) recently completed an interdisciplinary Ph.D. in environmental communication. Her research explores public participation, nuclear and risk communication, and environmental policy primarily within the Department of Energy's nuclear weapons complex, including discussions concerning environmental degradation and human health impacts of nuclear weapons production.

JUDITH HENDRY (Ph.D., University of Denver) teaches in the Department of Communication and Journalism at the University of New Mexico. Her work has focused on environmental rhetoric and public participation and has appeared in *Speaker and Gavel* and *Environmental Perspectives*. She is the author of the upcoming book *Contemporary Approaches to the Study of Environmental Communication*. She is the recent past president of the Environmental Communication Commission of the National Communication Association.

WILLIAM J. KINSELLA (Ph.D., Rutgers University) is an assistant professor of Communication at Lewis and Clark College, Portland, Oregon. His research interests include organizational communication, environmental communication, and the communication relationships among science, technology, and public policy. He serves on the Hanford Advisory Board, which advises the U. S. Department of Energy, the U. S. Environmental Protection

Agency, and the Washington State Department of Ecology on the cleanup effort at the nation's most contaminated nuclear site, and is vice-chair of the board's Public Involvement and Communication Committee.

PHAEDRA C. PEZZULLO (Ph.D., University of North Carolina) is an assistant professor in the Department of Communication and Culture at Indiana University. Her essay in this volume is derived from her dissertation work on the rhetorical performance of toxic tours. She previously has published on Warren County, NC, the birthplace of the environmental justice movement, and on commercial tourist practices in national parks.

STEVE SCHWARZE (Ph.D., University of Iowa) is an assistant professor in the Department of Communication Studies at the University of Montana. His research interests include environmental rhetoric, rhetoric and public policy, argumentation and classical and contemporary rhetorical theory. He is especially interested in how governing institutions justify environmental policies, and how citizens engage those institutions rhetorically. He is currently investigating controversies surrounding environmental and public health hazards left as a legacy of mining in Montana.

SUE L. SENECAH (Ph.D., University of Minnesota) is an associate professor in the Environmental Studies Department at the State University of New York College of Environmental Science and Forestry. Her research and teaching focuses on the design and evaluation of environmental communication and participatory processes for environmental decision making concerning, for example, watershed management, wildlife management, regional planning, hazardous waste remediation, and park/protected area management. Further, since 1992, she has served as the Special Assistant for Environmental Policy for a New York State Senator whose district includes Love Canal. Finally, Professor Senecah practices as a mediator, facilitator, and process design consultant for environmental public policy decision making.

AMOS TEVELOW (Ph.D. candidate, University of Pittsburgh) is an assistant professor in the Department of Speech Communication at Ithaca College in New York. His work analyzes nonprofit research organizations as strategic actors in the construction of domestic and global policy agendas. He also studies the content and political economy of news corporations, and urban and transnational social movements. His most recent publication,

"Consensus and Conflict: Social Capital and Community Reinvestment in Pittsburgh," is a coauthored study of how community-based organizations shape development agendas at the local level (*Proteus: A Journal of Ideas*, November 2001).

GREGG B. WALKER (Ph.D., University of Kansas) is professor and former chair of the Department of Speech Communication, adjunct professor of Forest Resources, and director of the Peace Studies program at Oregon State University in Corvallis. Over the past decade Walker has published a number of articles and presented numerous papers on environmental conflict management and dispute resolution. He is the coauthor of *Working Through Environmental Conflict: The Collaborative Learning Approach* (Praeger, 2001). Off campus, Walker conducts training programs on collaborative decision making, designs collaborative public participation processes, facilitates collaborative learning community workshops about natural resource and environmental policy issues, and researches community-level collaboration efforts.

CAITLIN WILLS TOKER (Ph.D., University of Georgia) is an instructor of Communication at Gainesville College in Gainesville, Georgia. Wills-Toker uses a critical rhetorical perspective to explore communication in the public sphere of environmental participation. She believes that the issues surrounding democratic participation are especially heightened in this context of conflict and technical sophistication.

Index

1992 Earth Rio Summit, 231
1998 Summit of the Americas, 209–10; recommendations for participation of civil society groups, 215
2001 Summit of the Americas, 211, 217

Adaptive Management Areas (AMA). *See* USDA-FS, Northwest Forest Plan
Administrative Procedures Act of 1946, 56, 211
Advisory Commission on Intergovernmental Relations, 32

ballot initiatives, 90–91
Boundary Waters Act (BWA), 144–5, 147, 151
Boundary Waters Canoe Area (BWCA) Citizens Task Force, 144–5
Boundary Waters Canoe Area Wilderness (BWCAW), 6, 138, 148
Bush, President George W., 211

Chester Residents Concerned for Quality Living, 235
Cispus Adaptive Management Area. *See* USDA-FS, Northwest Forest Plan
citizen review panel, as alternative to public hearing, 158. *See also* community advisory board
civic discovery, 159, 168; achieved by Fernald Citizens Task Force (FCTF), 170; not achieved by Fernald Health Effects Subcommittee (FHES), 170–71
Civil Rights Memorial, 242
civil society, 7, 201; defined, 203; future challenges, 214–17; threatened by economic globalization, 202, 205; weakened by trade agreements, 206. *see also* Free Trade Agreement of the Americas, North American Free Trade Agreement
Clean Air Act, 36, 119
Clean Water Act, 36, 119
Clinton, President William J., 43, 180, 211; Executive Order on Environmental Justice, 218; "Reinventing Government" initiative, 113–4, 181; statements on Roadless Initiative, 125–6
Coastal Environmental Organization (CEO), 183. *See also* Georgia Ports Authority
collaborative learning, 3; as part of collaborative decision making, 123; situation mapping as tool, 46–7
collaborative potential: defined, 123; versus traditional mechanisms of public

participation, 124; in Roadless Initiative case, 131 (*see also* USDA-FS, Roadless Initiative of 1999)
communication theory, 10
community advisory boards, limitations, 90. *See also* citizen review panel
community capacity, 13
competing values approach: applied to Fernald case, 70–78; defined, 63–64; implications, 79–81; linked to criteria for democratic public participation, 78–79; linked to expectations about public participation process, 63; perspectives (rational, political, consensual, empirical) defined, 64–65; possibility for blending of perspectives, 79
Comprehensive Environmental Response, Compensation, and Liability Act (CERCLA), 59, 60, 66–7, 168
COSMOS Corporation, 173. *See also* Fernald Health Effects Subcommittee, USDDH-CDC
Council on Environmental Quality (CEQ), study of National Environmental Policy Act (NEPA) implementation, 100
Cromer-Campbell, Tammy, 256, 258–60, 262. *See also* M. O. S. E. S.
cultural activism: art as fountainhead, 264–65; defined, 255–56, 265; *Fruit of the Orchard (FOTO)* photography exhibit as, 259–60, 263; offers alternative to traditional public participation, 255, 261–2; source of empowerment and community voice, 262–63; strategy for change and liberation, 259, 263–4. *See also* M. O. S. E. S.

decision space: and decision authority, 119–21; defined, 119; in Roadless Initiative case, 125–26. *See also* USDA-FS, Roadless Initiative of 1999
Dewey, John, 21
deliberative democratic theory. *See* public participation, consensus-based stakeholder approach
dialogic conception of public participation. *See* public participation, democratic model

Earth in the Balance, 244
Earth Justice Defense Legal Fund, 211
Endangered Species Act, 36
economic globalization, 201; critics of, 202; defined, 204; historical antecedents, 223; impact on civil society, 205
environmental communication: definition, 4; functions, 4–5; links to environmental problems and solutions, 112
environmental issues: conflicts, 14, 25; growing complexity, 15
environmental justice: challenges faced by communities, 240; defined as problem, 236, 240
European Union (EU), 215

fairness and competence, 3, 62–63. *See also* public participation, democratic models
Federal Advisory Committee Act (FACA), 211
Fernald Citizens Advisory Board (FCAB). *See* Fernald Citizens Task Force
Fernald Citizens Task Force, 159; access to information, 163–65; building interpersonal relationships with DOE, 71–2; formation of, 67, 161; "Future of Fernald" process,

171–72; Future Site game, 75; impact on DOE decision making, 76–78, 167–68; initial recommendations, 67, 170; leadership, 163; learning about Fernald site, 74–76; membership, 162; open and ongoing communication with DOE, 73–4, 166–67; overall effectiveness, 169–70; recommendations since 1997, 171; renamed Fernald Citizens Advisory Board (FCAB), 67, 171
Fernald dose-reconstruction study, 172
Fernald Health Effects Subcommittee, 6, 159; access to information, 164–65; degree of influence upon CDC and ATSDR, 168–69; face-to-face interaction with agency officials, 166–67; formation of, 161; improvements in meeting formats, 172; leadership, 163; membership, 163; overall effectiveness, 170–71; terminated, 172–73
Fernald Living History Project, 61, 68–70, 81
Fernald Residents for Environmental Safety and Health (FRESH): building interpersonal relationships with DOE, 71–2; cancer map, 75; description, 66; impact on DOE decision making, 76–78; initial attitude toward DOE, 70; learning about Fernald site, 74–76; open and ongoing communication with DOE, 73–4; role in FHES, 166
Fernald site. *See* USDOE-FEMP
Fischer, Frank, 84–5, 88–89, 91, 94
Fisher, Walter, 83, 86–88
Fluor Daniel Fernald, 67, 68. *See also* USDOE, Fernald site
Forest and Rangeland Renewable Resources Planning Act, 36
Forest Ecosystem Management Team (FEMAT), 39–40; report on public participation, 47–49
framing, 110; "public interest" frame, 110–111
Free Trade Agreement of the Americas (FTAA), 6, 7, 203; barriers to participation by civil society groups, 211–14; Committee on Government Representatives on the Participation of Civil Society (CGR), 209–11; negotiations, 209, 211; non-transparent dispute resolution process, 214
Fruit of the Orchard (FOTO), 258–60; as cultural activism, 259–60. *See also* cultural activism, M. O. S. E. S.
fundamental interpersonal relations orientation (FIRO) theory, 13, 22; application to public participation, 22

General Agreement on Tariffs and Trade (GATT), 202, 224; renegotiation in Uruguay round, 224–25
Georgia Department of Natural Resources, 188. *See also* Georgia Ports Authority
Georgia Environmental Protection Division, 184. *See also* Georgia Ports Authority
Georgia Ports Authority (GPA), 7, 176; formed, 181
GPA Harbor Deepening Project 177, 181–82; conducted project feasibility study granted by Water Resources Development Act (WRDA), 182–83; opposition by environmental groups, 183, 197; role of regulatory agencies, 183–84
GPA Modeling Technical Review Group (MTRG), 193–96, 197
GPA Stakeholder Evaluation Group (SEG), 176, 183; GPA control of

SEG agenda, 186–90; GPA definition of consensus, 190–93; GPA definition of stakeholder, 193–97; transformed from consensus-based group to technical advisory group, 197–98
Glazer, Phyllis, 259, 266; founded M. O. S. E. S., 257–58. *See also* M. O. S. E. S.
Government in the Sunshine Act, 211
Greenpeace Toxics Patrol, 248

Habermas, Jurgen: communicative competence, 92; instrumental vs. communicative rationality, 140, public sphere, 203

Interhemispheric Regional Workers Organization (ORIT), 216–17
International Association for Public Participation (IAP2) public participation spectrum, 117–9
International Monetary Fund, 202, 223; efforts to resolve 1997 financial crisis, 226; limited role of NGOs, 227

Jesus People Against Pollution, 247–48

ladder of citizen participation, 29
Las Placitas Association (LPA), 104, 108. *See also* USBLM, Placitas, New Mexico case
legitimacy, 5, 64, 137, 138; analyzed in Boundary Waters Canoe Area Wilderness (BWCAW) case, 143–151 (*see also* USDA-FS, BWCAW management plans); as legitimate rhetorical goal for agencies, 151–53; as rhetorical effect, 142–43; dimensions of, 142; global governance crisis, 223
Leopold, Aldo, 23
Louisiana's "cancer alley," 249

Metalclad Corporation v. Mexico, 206. *See also* North American Free Trade Agreement (NAFTA)
Methanex Corporation case, 207. *See also* North American Free Trade Agreement (NAFTA)
Mothers Organized to Stop Environmental Sins (M. O. S. E. S.), 256, 261; committed to activism, 263–64; example of cultural activism, 264; founded, 257–58; opposed siting of facility in Winona, Texas, 257–58; subject of *FOTO* exhibit, 258–60. *See also* Cromer-Campbell, Tammy; cultural activism; *FOTO*
Multiple-Use and Sustained Yield Act, 151

National Environmental Policy Act (NEPA), 1,15, 27, 35, 36, 99, 111, 113, 115, 151, 175; CEQ evaluation, 102–103; critique of Environmental Impact Statements, 140–2, 155; Environmental Assessment (EA), 101; Environmental Impact Statement (EIS), 102, 105, 182; Finding of No Significant Impact (FONSI), 101; handbook for implementation, 101, 105, 106; Mitigated FONSI, 102; No Action Alternative, 102, 106–7; public participation requirements, 6, 16, 36, 60, 99–100, 115–6, 218; use of Environmental Assessments as advocacy tool, 112
National Forest Management Act, 36, 126, 145, 151, 152–3
National Institute for Occupational Safety and Health (NIOSH), 172

Nature Conservancy, 183. *See also* Georgia Ports Authority (GPA)
neo-liberalism, 204–05, 227. *See also* economic globalization, "Washington consensus"
New York State Department of Environmental Conservation, 28
New York State Quality Review Act (SEQRA), 27, 28, 31; draft environmental impact statement (DEIS), 27, 28, 31; environmental impact statement (EIS), 27, 31
New York State Homebuilders Association, 30
non-governmental organizations (NGOs), 7, 16, 19, 224; role in IMF, 226–27; role in OECD, 224–26; role in UN, 230–31; role in World Bank, 227–30; role in WTO, 230; social capital in NGO networks, 231–32
North American Free Trade Agreement (NAFTA), 6, 7, 203, 205–6, 223; *Metalclad Corporation v. Mexico* case, 206–07; Methanex Corporation case, 207; non-tariff barrier authority under "Chapter 11" provision, 206; non-transparent dispute resolution process, 207–09, 214
Northwest Forest Plan. *See* USDA-FS
not-in-my-backyard (NIMBY) syndrome, 18

Office of the US Trade Representative, 209, 210
Ohio Environmental Protection Agency: attitudes toward DOE, 70; attitudes toward Fernald participation process, 71–73, 77–80; role in Fernald clean-up, 65
Organization of American States (OAS), 216
Organization for Economic Cooperation and Development (OECD), 225; effort to negotiate a Multilateral Agreement on Investment (MAI), 225; limited role of NGOs, 225–26

participatory democracy, 15. *See also* public participation, democratic models
Placitas, New Mexico. *See* USBLM, Placitas, New Mexico case
"post-Washington consensus," 226–27, 229–30, 231. *See also* economic globalization, "Washington consensus"
practical theory, 14–15; defined, 21–22, outcomes, 22; Trinity of Voice (TOV) approach as example, 33
Public Citizen, 211, 214
public expertise: alternative role for technical expert, 88–89, 94; cultivating, 92–94; defined, 84, 86–7; importance in enhancing legitimacy of decisions, 94–5; versus traditional role of technical expert, 83, 95
public hearings, 17–18; criticisms, 27–31; lack of assessment research, 20; proxemics, 28–9; role of technical experts, 89 90. *See also* public participation, traditional mechanisms
public moral argument, 83, 86–87
public participation: assessing, 19–20, defined, 115; exclusion or marginalization of voices, 139–40; expectations about, 18; instrumental rationality and, 140; limitations of, 9; spectrum, 117–18
public participation, consensus-based stakeholder approach: analysis of Georgia Ports Authority (GPA) case, 186–98 (*see also* GPA SEG); based

on deliberative democratic theory, 177; consensus as goal, 178; criticisms of, 178–79, 198–99; describes participants as stakeholders, 180; seen as solution to problems with traditional approaches, 176; used in environmental decision making, 179–81
public participation, democratic model: application of criteria in Fernald case, 162–73; based on two-way view of communication, 62; criteria for effective democratic participation, 61–62, 158–59; dialogic conception, 48
public participation, social constructionist model, 100. *See also* social communication perspective
public participation, traditional mechanisms: based on one-way view of communication, 36–40; command and control approach, 121, 126–31; comparisons with innovative mechanisms, 120; consultation versus collaboration, 121–22; criticisms, 2–3, 16–17, 18, 100–01, 135, 175–76; "decide-announce-defend" strategy, 100; objectification of process, 141, 144–45, 146–47; participation gap, 60–61, 89, 91–92
presence, 237; defined, 244–46, 253; not to be confused with reality, 246–47
public vocabulary: defined, 176, 184; elements of (ideographs, narratives, myths, characterizations), 185–6

Reinventing Government Initiative. *See* Clinton, President William J.
Resource Conservation and Recovery Act (RCRA), 67
risk, technical and cultural understandings, 60
risk communication: traditional approach, 157–58; dialectical approach, 158–59 (*see also* public participation, democratic models)
"risk society," 83
Roadless Area Directive. *See* USDA-FS, Roadless Initiative of 1999

Savannah National Wildlife Refuge, 183. *See also* Georgia Ports Authority
Seventh American Forest Congress, 132–3
Sierra Club, 248
social capital: defined, 224, 233; employed by United Nations, 230–31; employed by World Bank, 229–30; employed by World Trade Organization, 230; limitations, 231–32
social communication perspective: applied to Cispus Adaptive Management Area (AMA) case, 43–55; assumptions, 40–41; elements (openness, shared responsibility, interpersonal relationships), 41–43; evaluation of public participation programs, 55–56; issues of control and shared responsibility, 54–55
Southern Poverty Law Center, 242
stakeholder, 5, 14, 16, 30, 99, 104, 106, 171; conflicts, 83; defined in Georgia Ports Authority case, 193–97 (*see also* GPA SEG); place-based processes, 112; role in consensus-based public participation approach, 180, 186
Strategic Litigation Against Public Participation (SLAPP) suits, 19
Southern Environmental Law Center (SELC), 183. *See* Georgia Ports Authority (GPA)

technocracy, 15
Texas Natural Resources Conservation Commission (TNRCC), 257–58, 264
terministic screen, 154, 253
Toxic Release Inventory, 248–49
toxic tours: as communication strategy, 244; as environmental advocacy strategy, 236–7, 242–4; as means of considering modernity, 251–2; criticism of, 247; defined, 236; presence and, 244–47; purpose of, 247–50; related to concepts of tour and tourist, 241–2
toxics: scope of problem, 237–39; as symptom of modernity, 239–40
trinity of voice (TOV) approach, 5, 22; defined, 23; elements of (access, standing, influence), 23–5; examples of grammars (tactics, behaviors), 26–27; applied to public hearings, 27–31; relationship to trust-building, 31–32
trust, 20–21, 23, 32, 108; decline of public, 60; importance of interpersonal relationships in building, 71–72, 78

United Nations (UN), 2; efforts to build social capital among NGOs, 230–31; global governance efforts, 231
UN Commission on International Trade Law, 208
University of Cincinnati Center for Environmental Communication Studies, 68
US Bureau of Land Management (USBLM), 2, 36
USBLM, Placitas, New Mexico case: agency jurisdiction, 103; proposal from Western Mobile, 104; EA process, 105–6; threat of No Action Alternative, 106–8; Mitigated FONSI, 108–9; framing public good, 110
US Department of Agriculture-Forest Service (USDA-FS), 2, 6, 36, 39, 137; accountability, 151–2, 156; *Public Participation Handbook (1992)*, 36–38; public participation process, 36–40
USDA-FS, Boundary Waters Canoe Area Wilderness (BWCAW) management plans: definition, 143–4; 1981 Implementation Plan, 144–6; 1986 Superior National Forest Management Plan, 146–8, 149; 1993 Boundary Waters Management Plan, 148–9
USDA-FS, Northwest Forest Plan, Adaptive Management Areas (AMA): designation, 43; implementation through Records of Decision (ROD) and Standards and Guidelines, 47–48; public participation activities, 44
USDA-FS, Northwest Forest Plan, Cispus Adaptive Management Area (AMA): assessment 53–55; Landscape Analysis and Design (LAD) process, 55–56; public participation activities, 44 53
USDA-FS, Roadless Intitiative of 1999: assessment of collaborative potential, 131, 134–5; comparison of draft EIS, final EIS, and Record of Decision (ROD), 127–8; public participation strategy, 124, 129–31
US Department of Health and Human Services, Agency for Toxic Substances and Disease Registry (USDDH-ATSDR), 161, 168
US Department of Health and Human Services, Centers for Disease Control (USDDH-CDC): assisted in selecting

FHES membership, 163; conducted health studies at DOE sites, 161; established FHES, 161–2; evaluated health effects advisory process, 173; failed to support FHES, 170–71; funded Fernald risk assessment studies, 172; limited range of policy options, 168; organized FHES meetings, 166–67; provided FHES with information, 165; terminated FHES, 172–73. *See also* Fernald Health Effects Subcommittee (FHES)
US Department of Energy (USDOE), 2, 6, 59, 60; Office of Environmental Management (EM) definition of public participation, 117; public participation process, 89
USDOE, Environmental Management Advisory Board, 163
USDOE-FEMP: description of Fernald site, 60, 65, 159–60; environmental contamination, 66, 159–160; cleanup under CERCLA, 66–7; public meetings, 66; initial relationship with FRESH, 66–7; built interpersonal relationships with stakeholders, 71–2; engaged in open and ongoing communication, 73–4; role of stakeholder input in decision making, 76–78
US Environmental Protection Agency (USEPA), 2, 35, 59, 119; public participation process, 20, 89; role in GPA Harbor Deepening Project, 183
USEPA National Environmental Justice Advisory Council (NEJAC): Model Plan for Public Participation, 117–8; toxic tour taken by Environmental Justice Enforcement and Compliance Assurance Roundtable, 250
US Fish and Wildlife Service (USFWS), 36; assessment of public participation programs, 139; public participation requirements under NEPA, 116; role in GPA Harbor Deepening Project, 183
US Holocaust Memorial Museum, 242

"Washington consensus," 204, 223. *See also* economic globalization, "post-Washington consensus"
Water Resources Development Act (WRDA), 182
Western Mobile. *See* USBLM, Placitas, New Mexico case
Wilderness Act, 151, 152
World Bank (WB), 2, 6, 202, 215, 223, 227; Center for Settlement of Investment Disputes, 208; debt relief projects, 228–29; dispute resolution process, 208; efforts to develop social capital, 229–30; NGO involvement in project development, 227–29; Structural Adjustment Participatory Review Initiative (SAPR), 215; transparency of process, 229
World Trade Organization (WTO), 6, 202; dispute resolution process, 214; protests at 1999 meeting in Seattle, 215, 223; role in economic globalization, 205; role of NGOs, 230

49615303R00182

Made in the USA
San Bernardino, CA
30 May 2017